Molecular Pathology

Molecular Pathology

Edited by

Jonathan R. Salisbury

King's College School of Medicine and Dentistry
London, UK

Taylor & Francis
Publishers since 1798

UK   Taylor & Francis Ltd, 1 Gunpowder Square, London EC4A 3DE
USA   Taylor & Francis Inc., 1900 Frost Road, Suite 101, Bristol, PA 19007

British Library Cataloguing in Publication Data
A catalogue record for this book is available from the British Library
ISBN 0-7484-0571-2

Library of Congress Cataloging Publication Data are available

Cover design by Jim Wilkie

Typeset in Melior by Keyword Typesetting Services Limited

Printed in Great Britain by T. J. International, Padstow, Cornwall

Contents

Contributors

Michael E. Edmonds
King's Diabetic Centre
King's College Hospital
Denmark Hill
London SE5 8RX

Nicholas Finer
Centre for Obesity Research
Luton and Dunstable Hospital
Lewsey Road
Luton
Bedfordshire LU4 0DZ

Michel R.A. Lalloz
Department of Haematological Medicine
King's College Hospital
Denmark Hill
London SE5 8RX

D. Mark Layton
Department of Haematological Medicine
King's College School of Medicine and Dentistry
Bessemer Road
London SE5 9PJ

William J. Marshall
Department of Clinical Biochemistry
King's College School of Medicine and Dentistry
Bessemer Road
London SE5 9PJ

Jonathan R. Salisbury
Department of Histopathology
King's College School of Medicine and Dentistry
Bessemer Road
London SE5 9PJ

General Preface to the Series

The curriculum for higher education now presents most degree programmes as a collection of discrete packages or modules. The modules stand alone but, as a set, comprise a general programme of study. Usually around half of the modules taken by the undergraduate are compulsory and count as a core curriculum for the final degree. The arrangement has the advantage of flexibility. The range of options over and above the core curriculum allows the student to choose the best programme for his or her future.

Usually, the subject of the core curriculum, for example biochemistry, has a general textbook that covers the material at length. Smaller specialist volumes deal in depth with particular topics, for example photosynthesis or muscle contraction. The optional subjects in a modular system, however, are too many for the student to buy the general textbook for each and the small indepth titles generally do not cover sufficient material. The new series *Modules in Life Sciences* provides a selection of texts which can be used at the undergraduate level for subjects optional to the main programme of study. Each volume aims to cover the material at a depth suitable to a particular year of undergraduate study with an amount appropriate to a module, usually around one-quarter of the undergraduate year. The life sciences was chosen as the general subject area since it is here, more than most, that individual topics proliferate. For example, a student of biochemistry may take optional modules in physiology, microbiology, medical pathology and even mathematics.

Suggestions for new modules and comments on the present volume will always be welcomed and should be addressed to the series editor.

John Wrigglesworth, Series Editor
King's College, London

Acknowledgements

It is my pleasure to acknowledge the enthusiasm and expertise of the chapter authors who have produced clear and concise accounts of their subjects. I thank them all for their hard work.

Alex Dionysiou in the Department of Medical Photography and Illustration, King's College Hospital, London, skillfully prepared the computer graphics that illuminate the text.

I am grateful to my colleagues Jane Codd, Gill Howes and the late Professor Bill Whimster in the Department of Histopathology, King's College School of Medicine, London, who read early versions of some chapters and suggested improvements; and to Professor Gerard Brugal of the Université Joseph Fourier, Grenoble, who kindly shared his ideas on cell proliferation with members of our department. I also thank Jane Codd for providing some of the illustrations in Chapters 1 and 6.

Jonathan R. Salisbury
London

1 An Introduction to Molecular Pathology

Jonathan R. Salisbury

1.1 What is pathology?

Pathology is the study of disease (*pathos* is a Greek word meaning 'suffering'). We think of the science of pathology in four parts:

1. The aetiology or causes of a disease. We understand the aetiology of some diseases very well (e.g. bacterial and viral infections). A specific example would be human immunodeficiency virus (HIV) causing the acquired immunodeficiency syndrome (AIDS). The cause of other diseases is essentially unknown. These are the cryptogenic or idiopathic diseases; for example, some cases of cirrhosis of the liver are classed as cryptogenic.

2. The pathogenesis of a disease. This means the actual mechanism of the disease causation starting from the normal state, i.e. how the cause(s) bring about the changes that we recognise as the disease state.

3. The morphological changes of a disease. This is a description of the structural changes in the diseased tissues.

4. The functional consequences of a disease. This is the impairment that the disease process causes; it can vary from none (subclinical disease) to death of the organism.

Mankind has lived with disease for all of his existence. Archaeologists often find evidence of disease processes in excavated skeletons, and the first naked eye observations of the morphological changes caused by disease processes must have been made in prehistoric times. It was not until 1761, however, that Giovanni Morgagni, working at the University of Padua in Italy, realized that, by systematically recording his observations at postmortem examination, he could document the changes brought about in a body by disease (although others had done this in a piecemeal fashion before him).

With the invention, and subsequent sophistication, of the light microscope, scientists turned to studying normal and diseased biological tissues at a light-microscopic level. We credit to Rudolf Virchow, a German doctor, the concept of 'cellular pathology'—that an understanding of disease processes could come about by studying the changes at a cellular level—a principle expounded in his book of 1858.

Nowadays, pathologists use a variety of non-molecular techniques to examine organs, tissues, cells or subcellular organelles that have been removed from diseased animals or humans. The starting point is naked eye observation (morbid anatomy, macroscopic or anatomical pathology), including a postmortem examination of a body. Light microscopy is used to examine tissue sections (histopathology) or cell smears (cytopathology). Electron microscopy allows the study of the ultrastructure of diseased cells.

The bacteria and viruses that cause specific diseases can be cultured from samples taken from the diseased part (these are the disciplines of microbiology and virology). Body fluids such as blood or urine can be analysed chemically to look for variations of the contained substances from the normal range (chemical pathology or clinical biochemistry). Haematology and immunology are the disciplines that study the cells and proteins of the blood and the immune system, respectively. The enormous benefit that has come from these non-molecular pathological techniques cannot be overstated; their daily application is saving countless lives and relieving much human suffering.

1.2 Why is molecular pathology different?

Molecular pathology is the study of disease processes at the level of nucleic acids, using the techniques of molecular biology. The definition of a molecule is that it represents the smallest portion to which a substance can be reduced without losing its chemical identity. Molecular biology, however, is the term used to describe the scientific study of DNA and RNA (and the proteins that regulate them), which grew from a collaboration of biochemistry, biophysics, cell biology and genetics in the 1970s. Just as light microscopy, and subsequently electron microscopy, brought a much greater understanding to pathological processes, so molecular pathology is shedding new light on old diseases and enabling us to explain some diseases that were not previously understood.

The chapters that make up this book are of two kinds: those that seek to illuminate our current understanding of the molecular pathology of a specific disease or group of diseases (e.g. diabetes mellitus or the haemoglobinopathies) and those that review a specific aspect of molecular pathology (e.g. genetic polymorphism or molecular histopathology) in relationship to all disease. It must be stressed, however, that a comprehension of molecular pathology on its own does not lead to an understanding of pathology as a whole; we have tried to achieve this fusion of molecular pathology with other specialities in the individual chapters.

1.3 Molecular pathology techniques and terminology

As so much of molecular pathology is technique-driven, the following section on the terms and techniques used in molecular pathology introduces some of the methods that are referred to in the subsequent chapters.

1.3.1 *Autoradiography*

Autoradiography is a method for obtaining images that relies on the detection of intracellular substances, labelled by radioactive molecules, by an X-ray film. Commonly used radioactive isotopes are ^{32}P, ^{35}S and ^{3}H.

Autoradiography is also used to describe a technique for studying cell function that provides tissues with radioactively labelled metabolites. Cells which take up the metabolite can then be detected by covering sections in a photographic emulsion. The radioactivity creates silver grains in the emulsion which are deposited onto the cell. If radiolabelled thymidine or some other DNA component is used, the autoradiograph will demonstrate which cells in the tissue are actively dividing. Autoradiography can also be used, for example, to determine which organ is metabolising a drug, and how quickly.

1.3.2 *DNA extraction and purification*

The starting point for many molecular biology techniques is DNA (Box 1.1). Standard DNA extraction methods begin with a tissue disruption step, performed either by physically crushing the sample (e.g. in a pestle with a mortar) or by cutting 'thick' (20 μm) cryostat or paraffin sections. The samples are then placed in a digestion buffer (to digest proteins) containing the enzyme proteinase K (which causes the proteolytic inactivation of nucleases) and incubated at 50°C for about 12 h. The DNA is then extracted using phenol and

Box 1.1 **DNA and RNA**

- Deoxyribonucleic acid (DNA) is a linear molecular chain composed of the four deoxyribonucleotides (deoxyadenosine 5'-phosphate, deoxythymidine 5'-phosphate, deoxyguanosine 5'-phosphate, and deoxy-cytidine 5'-phosphate). Two antiparallel chains of DNA (known as double-stranded DNA), together with associated basic proteins, form the 'backbone' of each of the chromosomes (see p. 6).
- Nucleotides are the basic subunits of DNA or RNA (ribonucleic acid) and consist of a base, a sugar and a phosphate. The sugar is deoxyribose in DNA and ribose in RNA. Uracil is substituted for thymidine in RNA. RNA molecules are single-stranded.
- The sequence of DNA nucleotides along the length of a chromosome is fixed and specific and, in certain places along the length of the chromosomes (the loci), this sequence forms the genes. Genes are found amongst much more DNA that is not transcribed (transcription is the synthesis of RNA from a DNA sequence).
- The genes are divided along their length into alternating sequences called introns (which seem to be inert) and exons. The exons carry the sequences of bases that determine the amino acid sequence of the protein translated from the gene (translation is the process of mRNA-directed protein synthesis in the ribosomes, see p. 6). Generally, exons encode distinctive structural and functional protein domains. This suggests that introns were present early in the evolution of genes.
- RNA copies of genes are made in the cell nucleus by RNA polymerases. These primary transcripts then undergo several modifications known collectively as RNA processing. The primary transcripts are modified at both ends, by the addition of a cap at the 5' end (which serves to protect mRNA against enzymatic degradation) and a poly-A tail at the 3' end. The introns are removed by internal cleavage and the exons joined together to produce mRNA. This modification is called splicing and is carried out in the nucleus by small ribonucleoprotein particles.

chloroform, precipitated with alcohol, spooled out on a glass rod, washed, briefly air-dried, dissolved in storage buffer and stored at $-70°C$. High-quality DNA can now be extracted from very small amounts (less than 50 mg) of tissue. Similar methods are used in forensic science (see p. 18) when the sample from which the DNA is extracted can be very crude, although the quality (the length of the nucleic acids obtained) is compromised.

1.3.3 *DNA probes*

A gene probe, also called oligonucleotide or primer, is a short (usually about 20 bases), separately prepared and *labelled* length of single-stranded DNA that has a sequence of bases complementary to those in the sample of genetic material to be investigated. A DNA probe can be labelled radioactively (e.g. with ^{32}P) or non-isotopically with fluorescein or with an enzyme for subsequent identification (Figures 1.1 and 1.2).

The DNA probes that provide the labelled single-stranded DNA for hybridisation are usually either plasmid probes or oligonucleotide probes.

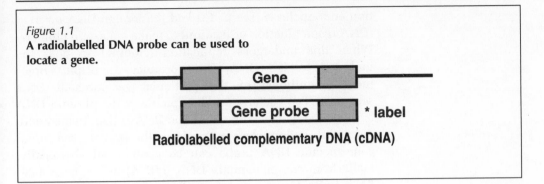

Figure 1.1
A radiolabelled DNA probe can be used to locate a gene.

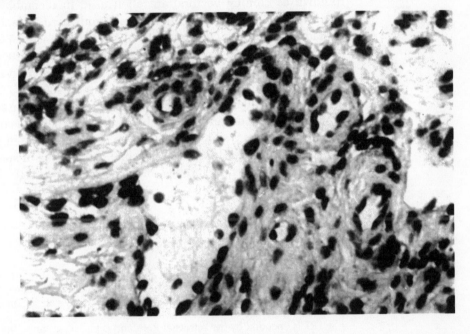

Figure 1.2
Pan-DNA in fibroblasts and endothelial cells in the connective tissues of the human uterine cervix. The probe is lengths of double-stranded, digested, genomic DNA labelled with fluorescein by nick translation. Detection is by anti-fluorescein antibody conjugated to alkaline phosphatase, visualized using nitroblue tetrazolium/5-bromo-4-chloro-3-indolyl phosphate. Courtesy of Jane Codd, King's College Hospital.

Plasmids occur in various bacteria and consist of an extra-chromosomal circular duplex of DNA that has the property of autonomous replication. They can encode for adhesion factors, antimicrobial (i.e. drug) resistance genes, bacteriocins and haemolysins. Because they can replicate inside bacterial cells, making numerous copies of themselves, they can be used to clone DNA. The plasmid is cleaved by a restriction endonuclease (see p. 13) and the foreign DNA inserted (DNA recombination or ligation) to make a plasmid chimera. When this undergoes replication (cloning) so does the included foreign DNA. This produces multiple copies (usually millions) of the particular recombinant DNA sequence. The bacteria are disrupted and the plasmid DNA is separated by centrifugation. The DNA is then treated with restriction endonucleases to release the probes. The initial gene-specific DNA probe can be synthesised chemically, made from complementary DNA (cDNA) using messenger RNA (mRNA) extracted from the relevant tissue and the enzyme reverse transcriptase (Box 1.2), or made from fragmented genomic DNA using restriction endonucleases. The plasmid probes are then labelled along the ribose phosphate backbone of the DNA and denatured to produce single-stranded DNA for hybridisation.

Plasmids vary considerably in size. Small plasmids (known as cloning vectors) are often used in recombinant DNA technology. Cosmids are large plasmids used as vectors to clone medium-length (40–50 kb) DNA sequences. Yeast artificial chromosomes are vectors, artificially constructed from the DNA sequences needed for replication in yeast cells, that can be used to clone large (up to 400 kb) DNA sequences. Bacteriophage vectors are also sometimes used.

Oligonucleotide probes are chemically synthesised short lengths of nucleic acid, often 14–40 bp (base pairs) in length. Oligonucleotide probes show the same high degree of specificity as long plasmid probes and have the advantage that they hybridise faster.

Box 1.2   **Reverse transcriptase (RT)**

- Reverse transcriptase (RT) is an enzyme, produced by certain RNA viruses (retroviruses), that synthesises DNA by copying an RNA sequence (the reverse of transcription, see p. 11).
- RT is encoded by the retroviral RNA and is packaged inside the viral capsid. Once inside a host cell, the RT makes DNA copies of the RNA genome. These DNA copies are then made circular or integrated into the host cell genome. Transcription of the integrated DNA by the host RNA polymerase then produces multiple copies of the viral RNA genome.
- RT can be used to obtain cDNA clones by producing complementary DNA copies (hence cDNA) from purified mRNA. These single-stranded cDNA molecules can then be converted into double-stranded DNA molecules and subsequently cloned (see p. 116).

1.3.4 DNA sequencing

This refers to the determination of the linear sequence of the nucleotides (see p. 4) in a given length of DNA. Sequencing procedures depend on the enzymes known as restriction endonucleases (see p. 13) and the application of powerful computer programs. We now know the complete DNA sequence of a number of microorganisms. These include free-living bacteria such as *Haemophilus influenzae* and *Mycoplasma genitalium*, the Archaea *Methanococcus jannaschii*, the blue-green algae *Cyanobacterium*, and brewer's yeast *Saccharomyces cerevisiae* (about 6000 genes).

Determining the 3 billion bp long sequence that makes up the human genome (a genome is the complete set of genes of an organism and the intervening DNA sequences) is the aim of a coordinated worldwide scientific endeavour known as the Human Genome Project (HGP). The HGP set itself three major scientific goals: creation of genetic maps, development of physical maps, and determination of the complete sequence of human DNA.

Comprehensive genetic maps of *Caenorhabditis elegans* (a 1 mm long nematode worm that is used extensively in the study of development as the adult worms contain only 959 cells), mouse and man have now been completed. The human genetic map is available as an extended reprint on the World Wide Web (http://www.genethon.fr). The complete human genomic sequence is projected to be completed by the year 2005.

1.3.5 Hybridisation

This is the artifical conjunction of two DNA strands, one of which usually carries either a radioactive or a non-radioactive marker. That the DNA strands are complementary means they can form a perfect double helix because they have a mirror-image relationship. Hybridisation is also used in a different sense to describe the production of cells containing chromosomes from more than one species.

1.3.6 Mutations: the basis of genetic variation

Mutations (variations in the sequence of base pairs) in DNA can be one of two basic types. Point mutations involve changes in a single base pair, whilst deletions, insertions and duplications involve a gain or loss of genetic material. Approximately 23% of point mutations are silent, either because the mutation does not alter the triplet genetic code (the code is degenerate, e.g. AAA and AAG both code for lysine) (polymorphism, see p. 156), or because the change in amino acid does not affect the structure or activity of the protein. A further 73% produce missense (cause an amino acid substitution which alters the structure or function of the protein), while approximately 4% are nonsense mutations, inducing termination of translation (i.e. they change the code to a stop codon). Since the genetic code is based on triplets of base pairs, insertion or deletion of genetic material involving numbers of bases not divisible by three (frameshift mutations) will result in the base sequence subsequent to the mutation losing its coding sense entirely from that point onwards, often with rapid degradation of the protein.

Chromosomal disorders are those in which there is a visible abnormality, detected by the discipline of cytogenetics, in the gross structure of a chromosome (e.g. trisomy or translocation). Many of these defects are not inherited, but some are. The presentation of chromosomal disorders is usually dominated by morphological abnormalities in the affected individual.

1.3.6.1 Patterns of inheritance

Man has 46 chromosomes, 44 of which are autosomes (common to males and females) and two of which are sex chromosomes. Females have two X chromosomes (one paternal and one maternal in origin), males one X and one Y chromosome (of maternal and paternal origin, respectively). There are 22 pairs of autosomes, one of each pair inherited from each parent. Each parent also provides one sex chromosome. As a result, we have two copies of each autosomal allele (the name given to the DNA sequence at a particular position, or locus, on a chromosome; an allele may be a gene but because much DNA appears not to code for proteins, not all alleles are genes or even parts of genes). If the two copies are identical, the individual is said to be homozygous for that allele; if dissimilar (i.e. one is a variant), the individual is heterozygous. Inherited diseases fall into one of three categories: autosomal dominant (only one copy of the mutant allele is required for the disease to be apparent); autosomal recessive (dose-dependent) (two copies are required), and sex-linked. The majority of sex-linked diseases are related to mutations in the X chromosome; these can be identified as there is *no* father-to-son inheritance. X-linked recessive disorders are usually only manifest in men since women also have a normal copy of the allele, and the abnormal copy is inactivated by lyonisation (Box 1.3). Women are car-

Box 1.3 **Lyonisation**

- Only one of the X chromosomes is genetically active.
- The other X chromosome, of either maternal or paternal origin, undergoes inactivation (with the exception of some genes at the tip of the p arm) and heteropyknosis to become the Barr body. This ensures that the amount of X-linked gene products in a female is equivalent to that produced in a male.
- Inactivation only occurs in somatic cells, since both X chromosomes need to remain active in the germ line.
- Inactivation of either the maternal or paternal X chromosome occurs at random among the cells of the blastocyst on or about day 16 of embryonic life.
- Inactivation of the same X chromosome persists in all the cells derived from each precursor cell.
- Because of lyonisation, X-linked traits in women are variably expressed. They may or may not be clinically expressed.

riers of the abnormal copy, however, and so can pass it to their offspring. There are around 300 X-linked recessive diseases in humans (e.g. colour blindness, fragile X syndrome, Duchenne and Becker muscular dystrophy). There are very few X-linked dominant conditions (e.g. hypophosphataemia—vitamin D-resistant rickets); they occur in men and women. Clinical expression is more constant and severe in hemizygous affected males than in heterozygous affected females in whom expression is variable due to lyonisation. Only a few inherited diseases due to mutations on the Y-chromosome are known; they can only occur in men, and pass from father to son.

In fact the situation is rather more complicated than this.

The concept of dominance is not straightforward, since homozygotes for a dominant condition are usually more severely affected clinically than heterozygotes; for example, heterozygotes with familial hypercholesterolaemia tend to have plasma cholesterol concentrations in the range 8–12 mmol l^{-1}, while in homozygotes the concentration can be of the order of 20 mmol l^{-1}. Both are prone to developing coronary heart disease but this tends to occur at an earlier age in homozygotes than in heterozygotes (see Chapter 5).

Furthermore, patients who appear to be homozygous for a mutant allele in that they express the phenotype (the observable consequences of the underlying genotype) of a recessive condition may actually be compound heterozygotes (a different mutation has been inherited from each parent). In females, one X chromosome in each cell becomes inactivated during embryogenesis (lyonisation, see Box 1.3). This inactivation is random unless there is loss of material from one X chromosome, in which case the structurally abnormal X is preferentially inactivated. However, lyonisation occasionally occurs in an atypical manner so that the mutant alleles are expressed; affected females are said to be manifesting heterozygotes. The severity of the disease is usually less than in a hemizygous male. Also, X-linked recessive diseases rarely can be observed clinically arising in women because of (1) Turner's syndrome (XO karyotype), (2) homozygosity when the gene is common in a population, and (3) X autosome translocation. Females with an X autosome translocation preferentially inactivate the normal X, otherwise the inactivation could spread from the inactivation centre in proximal Xq into the autosomal genes, leading to autosomal monosomy.

There are two further complications. It is common to find a degree of variation in the phenotypic expression of the same inherited disease in different individuals. There may also be variation in penetrance; that is, in whether the predicted phenotype is expressed at all in individuals whom genetic studies indicate must have the mutant genotype. Genotypic variation at other loci (i.e. multifactorial disease) and exogenous factors (e.g. diet) may be responsible for such apparent anomalies. Another factor involved in variation of expression is the nature of the underlying mutation. It is increasingly being recognized that many inherited diseases are intrinsically heterogeneous; that is, they can occur as a result of any one of several mutations affecting the gene in question. Since the effect of these mutations on the ultimate product of the gene may be very different, it is not surprising that the manifestations of the disease can vary also. Indeed, some inherited diseases that were thought to be distinct are now recognized to be due to abnormalities of the same gene product; this is true, for example, of Duchenne muscular dystrophy (DMD) (a severe condition) and Becker muscular dystrophy (BMD) (a milder condition), both of which are due to a variety of mutations (usually large deletions) affecting the dystrophin gene. Dystrophin is a membrane cytoskeletal protein that binds to actin and the sarcolemmal membrane and functions to maintain membrane integrity. In DMD, the deletions are nearly always 'out of frame' (a frameshift mutation, see p. 8) leading to virtual absence of protein. In BMD, the mutations are nearly always 'in frame' producing an interstitially deleted protein but still possessing amino- and carboxy-terminals.

A further factor which may underlie some variability of penetrance and expression is the phenomenon of imprinting; in fact, the expression of some genes seems to depend on whether they are of maternal or paternal origin, although this is probably relatively uncommon.

1.3.7 *Nucleic acid electrophoresis*

This is a method of separating fragments of nucleic acids by size using an electric field. A solution containing the fragments is placed onto a porous agarose gel and an electric potential difference applied between the two ends of the gel. DNA is negatively charged and so the DNA fragments move towards the anode, but this migration is hindered by the molecules of the gel. Larger fragments are hindered proportionately more, so the smallest fragments move furthest and become separated into bands. The DNA bands can then be visualised by staining with a molecule that fluoresces when illuminated with ultraviolet light (e.g. ethidium bromide).

1.3.8 *Polymerase chain reaction (PCR)*

The polymerase chain reaction (PCR) is a serial reaction that uses a thermostable DNA polymerase, isolated from thermophilic bacteria (e.g. *Taq* DNA polymerase from *Thermus aquaticus*, *Pfu* DNA polymerase from *Pyrococcus furiosus* or Vent DNA polymerase from *Thermococcus litoralis*), to amplify a DNA sequence millions of times in a few hours (Box 1.4).

The PCR technique starts with a heating step which denatures DNA isolated from cells into its complementary strands. These strands are then annealed (or hybridised) with two complementary oligonucleotides (see p. 4) (each about 15–20 nucleotides long) which are present in large excess. The two oligonucleotides (one for $5'$ sense and one for $3'$ antisense), which have been chemically synthesised to match sequences separated by n nucleotides where $50 > n < 2000$, serve as specific primers for in vitro DNA synthesis, catalysed by a thermostable DNA polymerase. This copies the DNA between the sequences corresponding to the two oligonucleotides. After multiple reaction cycles (i.e. denaturation, annealing, extension), typically 20–30 cycles, a large amount of the single DNA fragment, n nucleotides long, is produced. Trace amounts of RNA can be detected in a similar way by first converting them to DNA sequences with reverse transcriptase (see p. 6); this is called RT-PCR. Other variations (e.g. quantitation) also exist.

The development of PCR has produced a technological breakthrough in nucleic acid detection by increasing mole-

Box 1.4   **DNA polymerase**

- Discovered in 1957, DNA polymerase is the enzyme that creates the phosphate bonds between the deoxyribose sugars in the backbone of the DNA double helix, polymerising deoxyribonucleotide triphosphates on a single-stranded DNA template at the rate of about 50 nucleotides s^{-1} in mammalian cells.
- DNA polymerase is important for maintaining the integrity of DNA sequences and the stability of genes. It acts after DNA repair nucleases have removed damaged nucleotides, leaving a gap in the DNA helix. (The damaged DNA sequence must first be recognised as distinct from undamaged DNA; the damaged strand is then cleaved by an endonuclease, and an exonuclease then removes the damaged strand, leaving a single-stranded gap.) The DNA polymerase copies the information in the undamaged strand, one nucleotide at a time.
- DNA polymerase is vital for DNA replication prior to cell division, and is required for DNA templating (the process whereby the nucleotide sequence of DNA is copied by complementary base-pairing into a complementary nucleic acid sequence). During replication, each existing DNA strand serves as a template for the formation of a new DNA strand. Each of the two daughters of a dividing cell then inherits a DNA double helix formed from one old and one new strand.

cular sensitivity. This means that a particular segment of DNA can be amplified without cloning. The PCR technique is well known for its ability to amplify even single copies of DNA in a sample, and this includes contaminants. PCR must therefore be performed with meticulous attention to cleanliness, including a dedicated set of pipettes and preparation of the PCR mixture away from the amplification area (preferably in a different room).

1.3.8.1   PCR in microbiological diagnosis

There is increasing use of PCR methodology for bacterial diagnosis such as *Mycobacterium*, *Leishmania*, *Tropheryma whipplelii* in Whipple's disease, and for human papillomavirus subtypes in cervical cytology specimens by PCR on additional samples. It is well established, for example, that *Mycobacterium tuberculosis* can be found in tissues by PCR when Ziehl–Neelsen histochemical or auramine/ rhodamine fluorescence staining for acid-fast bacilli is negative. This is important because about 40% of culture-positive cases are negative on acid-fast bacilli staining and, as culture takes several weeks, specific antituberculous therapy may be delayed. PCR-based assays provide a rapid and reliable method for the detection of mycobacteria and hence enable early therapeutic intervention. Currently, these tests are being done only in specialised laboratories, but, with the production of

Figure 1.3
Human papillomavirus subtypes 16 and 18 demonstrated in the metaplastic stratified squamous epithelium of the transformation zone of the uterine cervix. The probes are oligonucleotides of 30 bases labelled with fluorescein and detected using an antifluorescein antibody conjugated to alkaline phosphatase, visualized using nitroblue tetrazolium/5-bromo-4-chloro-3-indolyl phosphate. Courtesy of Jane Codd, King's College Hospital.

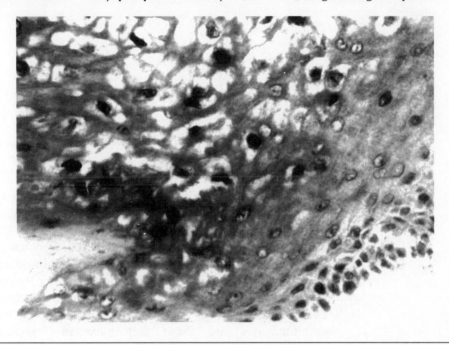

commercial 'kits', their use may spread to routine diagnostic laboratories.

1.3.9 *Restriction endonucleases*

These are enzymes which cleave DNA at sequence-specific sites along its length to produce a number of short fragments (Figure 1.4). They are called endonucleases because they cleave DNA within the molecule as opposed to at the end (exonucleases).

Most restriction endonucleases are found in bacteria, a few are from algae. They are a means by which a particular microorganism protects its DNA from violation by foreign DNA. Their activity is restricted to 'foreign' DNA because host DNA is protected by methylation at the sequence-specific sites. Each enzyme is designated by the first initial of the genus followed by the first two initials of the species from which it was derived. These are followed by a strain

Figure 1.4
Restriction enzymes. These are bacterial enzymes that recognise short sequences of DNA and cut the DNA at that point (arrows). The pieces of DNA produced are known as restriction fragment lengths.

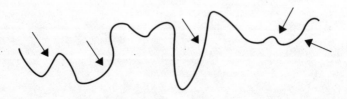

designation (if strain-specific) and a Roman numeral to indicate the order of discovery in that strain (Table 1.1).

Restriction endonucleases recognise specific sequences of 4–8 bases. On average, a given 4 bp site occurs every 256 bp, and a 6 bp site every 4096 bp. However, the frequency with which DNA is cut is affected by its base composition; for example, *Not*I has an 8 base recognition sequence that occurs very infrequently in mammalian DNA because it includes CG dinucleotides. In general, the most useful restriction endonucleases are those that have rare recognition

Table 1.1 Examples of restriction endonucleases

Enzyme	Microorganism	Sequence recognition site $5' \rightarrow 3'$
*Eco*RI	*Escherichia coli* RY13	G AATT C C TTAA G
*Hae*III	*Haemophilus aegyptius*	GG CC CC GG
*Hind*II	*Haemophilus influenzae* strain d	GTPy PuAC CAPu PyTG
*Not*I	*Nocardia otitidis-caviarum*	GC GGC CGC CG CCG GCG
*Taq*I	*Thermus aquaticus*	T CG A A GC T

Pu can be any purine base, Py can be any pyrimidine base.
Note that some enzymes produce blunt ends whereas others produce sticky ends.

sequences and therefore produce small numbers of fragments. *Not*I produces very large DNA fragments (1–1.5 million bp in length) that are extremely effective for physical mapping of DNA.

1.3.10 *Restriction fragment length polymorphisms (RFLPs)*

Restriction fragment length polymorphisms (RFLPs, often pronounced 'riflips') are the occurrence of two or more alleles (see p. 8) in a population, differing in the lengths of fragments produced by a restriction endonuclease (see p. 13) (Figure 1.5). Although polymorphism has a strict genetic definition (it represents the occurrence in a population of two or more genetically determined forms in such frequencies that the rarest of them could not be maintained by mutation alone), it has become used in various distinct senses. In RFLPs, polymorphism is used to imply alternative forms and, usually, that the commonest allele is less than 99%, so that over 2% of individuals are heterozygous.

RFLPs occur as a result of base changes, deletions, insertions and rearrangements which either create or destroy restriction enzyme cleavage sites. RFLPs have become of immense value in molecular medicine, as is discussed fully in Chapter 8.

1.3.11 *RNA extraction and purification*

There are a number of techniques for extraction of RNA (Box 1.5). One of the simplest and most effective methods is that of Chomzynski and Saachi, which uses a single extraction with an acid guanidinium thiocyanate/phenol/chloroform mixture. This provides a pure preparation of undegraded RNA in high yield and takes about 4 h. It is particularly useful for isolating RNA from minute quantities of cells or tissue samples. Some companies now produce commercial 'kits' for RNA extraction. These can be a practical alternative in certain circumstances but are relatively expensive. RNA is very labile in the presence of ribonucleases (RNases) which are released from the hands of laboratory workers and from damaged samples. Great attention has to be given to preventing RNase contamination of RNA samples.

The quality of extracted RNA can be assessed by denaturing electrophoresis and the observation of intact 28S and 18S ribosomal RNA. Should it be required to quantify the mRNA levels within a sample, this can be achieved, after electrophoresis, by transferring the RNA to a nylon membrane (see

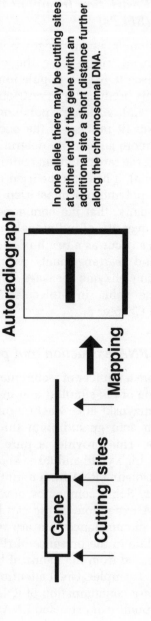

Figure 1.5
Identification of restriction fragment length polymorphisms.

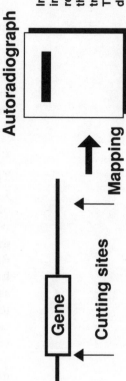

Box 1.5 **RNA species**

- RNA is found in three forms: messenger RNA (mRNA, about 4% of cellular RNA), which carries the genetic message from the genes to the ribosomes, ribosomal RNA (rRNA, about 85% of cellular RNA) which forms the RNA component of the ribosomes; and transfer RNA (tRNA, about 10% of cellular RNA), which are the small RNA molecules to which amino acids attach and which act as adaptors aligning these amino acids on the mRNA templates in response to the encoded genetic instructions.
- RNA is almost always single-stranded and linear (although a few circular RNAs are known to occur. The circular ones are the viroids of plants in which extensive base-pairing allows the formation of stiff, double helix-like structures).
- Almost all RNA molecules have many short double-helical regions which can form because two sections of the RNA chain within a hairpin fold are in the correct anti-parallel orientation to base-pair.
- tRNA species have evolved to contain extensive internal complementary sequences that base-pair to give virtually all tRNAs a specific tertiary shape (rather like an L), despite differences in size and secondary structure. This three-dimensional structure has sufficient conformational flexibility to satisfy the various structural demands to which tRNA is subjected. tRNAs are formed from larger precursor transcripts by processing nucleases, and then subjected to modifying enzymes that introduce a variety of different base or backbone substitutions. The tRNAs are then ligated to their cognate amino acids by aminoacyl-tRNA synthetases. The amino-acylated tRNAs transport their amino acids to the ribosome and their tRNA anticodons interact with complementary codons in the mRNA to deliver their amino acids to the growing polypeptide chain. The tRNAs are then recycled.
- rRNA molecules, together with proteins, make up the spherical ribosomes on which protein synthesis occurs. All ribosomes are constructed from two unequal subunits, one large and one small, known as the 28S and 18S rRNA subunits, respectively. In mammals, the large subunit contains 32 different proteins and the small subunit contains 27 different proteins. Ribosomal construction occurs in the nucleolus by importing the ribosomal proteins from the cytoplasm, assembling them with the rRNA molecules, and then re-exporting the completed ribosomes back into the cytoplasm. The genes for rRNA (rDNA) occur at discrete chromosomal areas called nucleolar organiser regions (see p. 113), the regions at which nucleoli form. In man, the rDNA units are found at five chromosomal loci (the acrocentric stalks of chromosomes 13, 14, 15, 21 and 22).
- In almost all circumstances, a single protein is translated from a single mRNA. Some exceptions are known to occur in *Escherichia coli* where some single proteins (which appear destined for degradation) are translated from two separate RNAs, with a 363-nucleotide RNA acting both as a tRNA (for alanine) and as mRNA.

Northern blotting p. 18) and then probing the membrane for constitutively expressed mRNA, such as that encoding glyceraldehyde-3-phosphate dehydrogenase (GAPDH). GAPDH is a key enzyme in glycolysis, the essential metabolic pathway by which cells begin converting glucose to energy, and is also a uracil-DNA glycosylase. Other functions of GAPDH include binding actin, ATP, amyloid precursor protein, calcyclin, RNA and tubulin.

1.3.12 *Southern, Northern and Western blotting*

Southern blotting allows for the transfer of DNA fragments from a gel to a nitrocellulose filter (see Southern, 1975). The gel containing the DNA is rinsed and placed on a filter paper wick in a bath containing sodium chloride/sodium citrate buffer solution. A nitrocellulose filter is put on top of the gel, followed by several layers of dry filter paper held down by a weight. The buffer is drawn by the dry filter paper through the gel, carrying with it the DNA fragments. When the DNA comes into contact with the nitrocellulose filter, it binds to it strongly (the DNA can be stored indefinitely in this form). The fragments can be permanently fixed to the filter by baking it in an oven at 80°C in a vacuum.

Northern blotting is a non-eponymous variation of Southern blotting relating to the transfer of RNA, rather than DNA, to a backing sheet prior to hybridisation.

Western blotting is a method for transferring proteins, separated by electrophoresis, to a membrane. It is analogous to Southern blotting for DNA or Northern blotting for RNA.

1.3.13 Gene mapping

Figure 1.6 illustrates how various techniques described in this section are used to isolate and identify a length of DNA containing a gene of interest.

1.4 Forensic science and molecular pathology

Forensic science is the discipline devoted to the scientific study of evidence, particularly evidence related to a crime. Forensic scientists were quick to realise the potential of DNA technology and to develop formidable new methods for examining biological material.

1.4.1 DNA profiling (DNA fingerprinting)

DNA profiling (a preferred term to DNA fingerprinting) was developed in 1984 by Professor Alec Jeffries, a geneticist at Leicester University. A DNA profile is a pattern of DNA sequences, such as tandem repeat sequences (multiple copies of a short DNA sequence that lie in series along a chromosome), the number of which is virtually unique to an individual. DNA is extracted from the initial sample and then cut with restriction endonucleases (see p. 13) to produce short segments. These are separated by nucleic acid electrophoresis (see p. 11) and then hybridised (see p. 7) with radioactive DNA probes complementary to the sequences of DNA repeats. The now-labelled sequences of

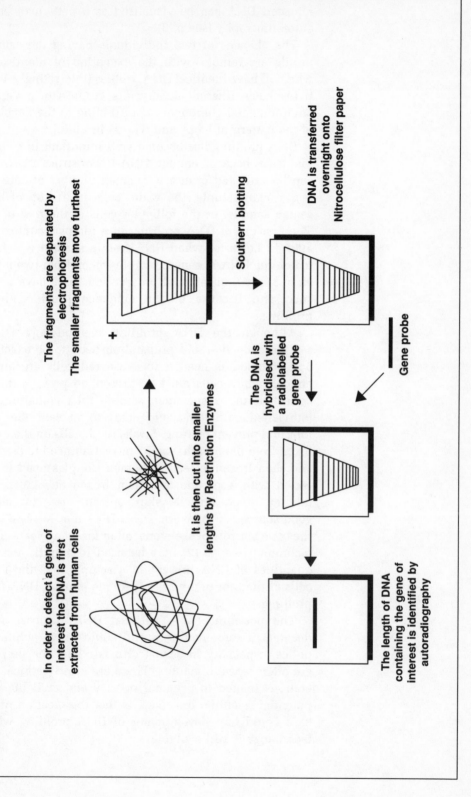

Figure 1.6
Gene mapping.

repeated DNA can be visualised as a pattern of bands by autoradiography (see p. 3).

The chances of two individuals having the same DNA profile are remote, with the exception of identical twins who will have identical DNA profiles (interestingly, identical twins have different fingerprints because of a very small environmental component contributing to the development of the pattern of loops and whorls in utero).

DNA profiling has become very important in some criminal cases because enough DNA to construct a DNA profile can be extracted from a very small number of cells indeed (e.g. a skin sample, the white cells in a drop of blood, a semen sample, or the follicular cells at the root of a hair). The first use of DNA profiling in a forensic context was in 1987 in Leicester when DNA from a suspect was found to match the DNA in semen samples taken from two murdered girls. Many thousands of prosecution cases have now been successful because of the evidence provided by DNA profiling.

Although the DNA profiling methodology is robust, sample collection and preparation need to be exact. About $5 \mu l$ of blood is needed (because red cells are anucleate, enough white cells must be present to provide the DNA), collected into a specimen sample tube containing EDTA (ethylenediaminetetraacetic acid) to chelate the calcium ions and prevent clotting. Swabs (with cells on their surface) taken from the mouth, vagina or rectum need to be air-dried and then frozen at $-20°C$. Semen samples need to be collected with a pipette and then frozen at $-20°C$. Tissues (kidney, liver, muscle, skin, spleen, etc.) are obviously good sources of cadaveric DNA. DNA may survive intact in the bone marrow of skeletons, often for many years. The PCR technique (see p. 12) may be used to amplify the minute quantities of DNA present in a sample (e.g. the follicular cells at the root of a hair) to provide enough DNA for DNA profiling.

The possibility of a national DNA database is being debated in some countries and could prove helpful to the police, especially in sexual crimes where the perpetrators are often repeat offenders. There are clearly ethical considerations related to personal privacy and civil liberty. One potential scientific drawback is that the creation of a database could halt development of DNA profiling while the technology is still evolving.

Further reading

Ausubel, F.M., Brent, R., Kingston, R.E., Moore, D.D., Seideman, J.G., Smith, J.A. and Struhl, K., 1987. *Current Protocols in Molecular Biology*. New York: Wiley-Interscience.

Brown, T.A. (Ed.). 1991, *Essential Molecular Biology: A Practical Approach*. Oxford: IRL Press.

Chomzynski, P. and Saachi, N., 1987. Single step method of RNA isolation by guanidinium thiocyanate–phenol–chloroform extraction, *Analytical Biochemistry*, **162**, 156–159.

Innis, M.A., Gelfand, D.H., Sninsky, J.J. and White, T.J. (Eds.), 1990. *PCR Protocols: A Guide to Methods and Applications*. New York: Academic Press.

Lachman, D.S. (Ed.), 1995. *PCR Applications in Pathology: Principles and Practice*. Oxford: Oxford University Press.

McPherson, M.J., Quirke, P. and Taylor, G.R. (Eds.), 1991, *PCR 1: A Practical Approach*. Oxford: IRL Press.

McPherson, M.J. and Hames, B.D. (Eds.), 1995. *PCR 2: A Practical Approach*. Oxford: IRL Press.

Miller, H., 1995. *Traces of Guilt. Forensic Science and the Fight Against Crime*. London: BBC Books.

Southern, E.M., 1975. Detection of specific sequences among DNA fragments separated by gel electrophoresis. *Journal of Molecular Biology*, **98**, 503–517.

2 Energy Metabolism and Obesity

Nicholas Finer

2.1 Introduction

In common with all living organisms, man requires energy. Human biology drives us to assimilate nutrients containing energy stored in the form of protein, fat, carbohydrates and alcohol (feeding). Ingested nutrients are processed to provide compounds that provide energy for immediate needs, and for energy stores that can provide a continuing supply of energy during periods of high energy requirements (e.g. exercise) or food shortage. Energy stores are essential to any living organism if it is to survive periods of food (energy) scarcity, or episodes of increased and frequently unpredictable energy demands (movement, pregnancy, illness). Human evolution has required that man has behavioural drives, and physiological and metabolic mechanisms that favour energy storage. Adipose tissue is the organ that serves this function. An ability to store energy efficiently and adequately, is advantageous to the survival of the individual and necessary for successful reproduction. It seems likely that the mechanisms favouring fat storage will have been genetically amplified by evolution. In very recent years, societal conditions in the developed world have altered so that energy expenditure has decreased and food availability increased. These factors have resulted in individuals, genetically adapted to a harsher environment, being prone to excessive fat storage and weight gain. Obesity is the term that describes this state in which energy stores (as fat) have become excessive to the point of disadvantage to health. Obesity is both a risk factor for many diseases, including coronary heart disease and diabetes, as well as a disease in its own right. It now affects 15% of the adult population in the UK and its prevalence is increasing year by year. An understanding of the mechanisms involved in human energy metabolism and fat deposition is perhaps one of the most pressing public health issues for the end of the millennium.

2.1.1 *Energy metabolism—basic principles*

Metabolism concerns the chemical processes involved in the growth and development of the body tissues, the maintenance of their integrity, the elimination of waste and breakdown products that result from this, and the energy involved in the functioning of the tissues. These processes require energy, which is derived from the energy contained in foodstuffs that are eaten. The metabolic processes that liberate energy (catabolism) in general are oxidative, and result in the formation of high-energy compounds which are then available for anabolic synthetic processes or functional processes (such as muscular contraction, maintaining ionic equilibrium across cell membranes, nerve conduction, etc.). Food (energy) intake, as well as metabolic demand, is periodic. Both components may vary over seconds or minutes (e.g. within meal feeding or muscular contraction), over hours or days (between meals, activity), or even longer term (famine and growth and development), and a variety of metabolic and storage processes exist to meet these needs. The study of energy metabolism seeks to account for the components making up total energy expenditure, as well as the sources providing, and the tissues using energy.

2.2 Components of energy expenditure

Energy expenditure has three main components: energy expended at rest (resting metabolic rate), energy expended at rest over and above resting expenditure (thermogenesis), and energy expended in activity and exercise (physical activity) (Figure 2.1). The qualitatively most significant component is basal or resting energy expenditure, a fact that is often found by patients and doctors alike to be surprising. Resting energy expenditure is, by definition, relatively stable on a day-to-day basis while thermogenesis and physical activity fluctuate considerably within and between days.

2.2.1 *Basal and resting metabolic rates*

The rate of energy utilisation at rest, after the effects of exercise have subsided and after the active absorption of ingested food has ceased, reflects the body's basal requirements. This basal metabolism reflects the costs of maintaining cellular integrity (e.g. Na^+/K^+ exchange across cell membranes) as well as whole organ activity (liver and brain activity, muscular work to maintain posture). In one sense it can be con-

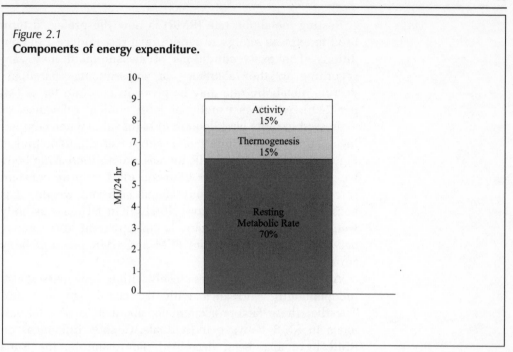

Figure 2.1
Components of energy expenditure.

sidered the minimum energy expenditure to maintain the body alive. The term basal metabolic rate (BMR) represents a now-outdated term for energy expenditure measured under stringent conditions of thermoneutrality, physical rest, and specified for time of day and nutritional state (after 12–18 h fast). Measurements made over a limited period are usually extrapolated to express BMR as an energy expenditure rate per 24 h (either in kcal or MJ). In common with many other physiological variables (e.g. renal glomerular filtration rate), attempts were historically made to normalise BMR measurements between individuals (and species) by expressing BMR in relation to body weight or, more commonly, body surface area. Such a practice is misleading and for many years lay behind the mistaken belief that the obese had low metabolic rates. Lean body mass or fat-free mass (FFM) has a disproportionate influence on BMR, since fat cells have a very low metabolic activity, both absolutely and in comparison to bone, muscle, nerve and other cells. Since the composition of weight gain is part fat (75%) and part fat-free mass (25%), it is clear that expressing BMR as energy expended per kilogram body weight or energy expended per unit surface area will suggest erroneously that the obese are metabolically more efficient than the lean. It cannot be emphasised too strongly that the obese have absolutely higher BMRs than the lean.

Resting metabolic rate (RMR) is now the preferred term used to express energy expenditure at rest under basal conditions. The exact conditions of measurement may vary according to the laboratory or experimental paradigm. Resting metabolic rate may be given in absolute terms (MJ per 24 h), which is usually of most clinical relevance, or expressed in units per kilogram FFM (of value when comparing groups of individuals or a subject at different times). Figure 2.2 shows how RMR increases with increasing body weight, and that RMR per kilogram FFM is quite constant across a wide range of body shape, size and weight. It is because FFM increases more slowly than fat mass as body weight increases that there is an apparent discrepancy between RMR per kilogram FFM and BMR per unit body surface area.

Although RMR correlates closely with fat-free mass, it also independently correlates with fat mass, age and sex. Together these factors account for about 80% of total variance in RMR between individuals. Genetic influences on RMR have also been described. For example, the Pima Indians from Arizona are a race with a predisposition to develop obesity (and diabetes mellitus); those with a low RMR seem more likely to become obese. In addition, a number of physiological influences on RMR have also been determined (Table 2.1).

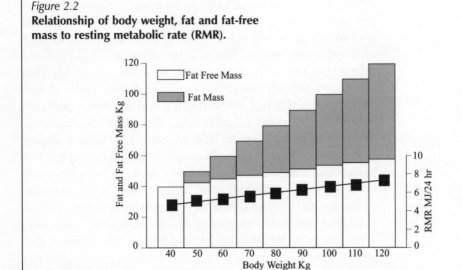

Figure 2.2
Relationship of body weight, fat and fat-free mass to resting metabolic rate (RMR).

Table 2.1 Physiological influences on resting energy expenditure

Resting metabolic rate per kg fat-free mass:	
Increased	Decreased
• Luteal phase of menstrual cycle	• Circadian rhythm—early morning
• Elevated catecholamines	• Sleep
• Fever	• Television viewing
• Thyrotoxicosis	• Underfeeding
• Injury	• Genetics
• Sepsis	
• Stress, mental activity	
• Sensory stimulation (e.g. smell)	
• Genetics	

2.2.2 Meal-induced thermogenesis

Food intake is one of the most important stimuli to increased metabolic rate. Food intake imposes obligatory energy expenditure for digestion, absorption, metabolic processing and storing of the ingested food. The energy costs of these processes will depend largely upon the metabolic pathways through which the nutrients are metabolised. Glucose oxidation is carried out with an efficiency of about 99%, while, by contrast, 5% of the energy of glucose is wasted in glycogen synthesis and 22% when it is converted by lipogenesis to lipid. The term thermogenesis is used to describe this obligatory energy expenditure since it is energy that is 'wasted' as heat. The term is slightly confusing, however, since this metabolic activity is commonly assessed by measuring increased oxygen consumption used for the additional aerobic metabolism. Much research has focused on whether defects in thermogenesis could predispose to weight gain, and a majority of studies have found reduced meal-induced thermogenesis in established obesity. Even in the positive studies, however, the size of the reduced meal-induced thermogenesis is small. At about 3–5% less than lean control subjects this reduction could, if energy intake remained constant, account for a weight gain of about 5 kg per year. However, since both energy intake from food and energy expenditure from activity vary daily to a much greater degree both within and between individuals, it seems unlikely that a minor reduction in one of the smaller components of total energy expenditure can be an important cause for obesity.

2.2.3　*Methods for measuring energy expenditure*

A number of techniques have been developed that will directly or indirectly measure energy consumption in human subjects. There always exists a compromise between the accuracy and completeness of measurements with the constraints on normal activities that measuring imposes.

2.2.3.1　*Direct calorimetry*

This technique measures the rate at which heat is lost from the body to the environment, by radiation, convection and conduction, as well as from the evaporation of water. The subject is placed within a chamber, often a small room allowing some physical movement. In one system a well-insulated chamber is used, and the rate at which heat must be extracted to maintain thermoneutrality measured (heat-sink calorimetry). An alternative is to use a poorly insulated chamber in which the temperature gradient across the wall of the chamber is measured. In both cases, evaporative heat losses are measured separately. The subject can spend protracted periods in the chamber to allow experiments involving dietary or exercise manipulations. The cost of these chambers is high and only a handful have been built worldwide.

2.2.3.2　*Indirect calorimetry: chambers*

Under normal physiological conditions, neither oxygen nor carbon dioxide is stored within the body. The rates of oxygen consumption and carbon dioxide production therefore measure the rate of metabolism of the major substrates (carbohydrate, fat, protein and alcohol). Hydrogen and methane exchange as a measure of fermentation energy-providing reactions can also be measured. In a whole-body indirect calorimeter, the subject is enclosed within a chamber and changes in the oxygen and carbon dioxide concentrations of air entering and leaving the chamber are measured. There are a number of methods to calculate energy expenditure from oxygen uptake, carbon dioxide production and flow. It is necessary to adjust for the volume difference between incoming and outgoing air which results from the different volume ratios of oxygen to carbon dioxide. Haldane, and later McLean, developed the following equations:

$$V_{O_2} = V_E[D_{fO_2} + 0.2561(D_{fO_2} + D_{fCO_2})]$$
$$V_{CO_2} = V_E[D_{fCO_2}]$$

where V_{O_2} is oxygen uptake ($l\,min^{-1}$), V_E is flow of expired air ($l\,min^{-1}$), D_f is change in fractional composition of the gas, V_{CO_2} is expired CO_2 ($l\,min^{-1}$).

A simplified calculation by Weir, in 1949, relates energy production to the volume of expired air and the difference in oxygen consumption between inspired and expired air:

$$E = 3.43 \times V_E \times D_{fO_2}$$

where E is the energy (watts).

This implies that 1 litre of oxygen is equivalent to 20.5 kJ, whatever the substrate oxidised, and regardless of the respiratory quotient (RQ). This formula works well in practice because two errors (the fall in energy equivalence of oxygen with falling RQ, and the relative volumes of expired to inspired air) effectively cancel each other out.

Figure 2.3 illustrates recordings taken from a calorimeter over a 24 h period in a lean and an obese subject.

Indirect calorimetry allows an assessment of the relative contribution of different substrates to total energy expenditure. The ratio of carbon dioxide production to oxygen consumption (the respiratory quotient, RQ) differs for each of the four substrates (Table 2.2). Protein and alcohol oxidation can

Figure 2.3
Schematic example of recordings from a calorimeter over a 24 h period in an obese and a lean subject.

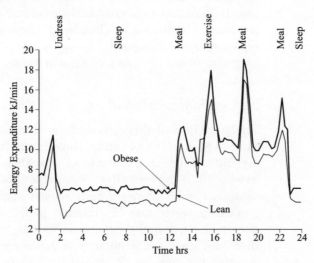

Table 2.2 Constant used in indirect calorimetry

	Fat	Carbohydrate	Protein	Alcohol
Respiratory quotient	0.71	1.00	0.82	0.66
Energy density (kJ g^{-1})	39.4	15.8	18.6	29.7
Energy density (kcal g^{-1})	9.4	3.8	4.4	7.0
Energy l^{-1} O_2 (kJ)	19.6	21.3	19.2	20.1
Energy l^{-1} O_2 (kcal)	4.7	5.1	4.6	4.8

measured by other techniques (urinary nitrogen excretion and breath alcohol disappearance) or allowed for as a constant. If these substrates are removed from the calculations, a non-protein and non-alcohol RQ can be calculated. This will be determined by the ratio of fat to carbohydrate oxidation, absolute amounts of which can then be calculated with a high degree of accuracy (about 10–20 g per day respectively).

2.2.3.3 Indirect calorimetry: free-living conditions

Rather than collecting and measuring expired air from a chamber, the expired air can be collected directly through a mouthpiece into a spirometer or bag (Douglas bag), or sampled from air drawn into a hood enclosing the subject's head (ventilated hood). The expired air is collected over a period of time. The volume (either absolute, or from flow rate and time) and concentration of gas relative to inspired air are measured. Ventilated hoods avoid the need to ensure a tight fit between the mouth, nose and tubing into the collecting bags (Figure 2.4). Since residual air volumes in these systems are small (compared to chambers), these systems are very responsive to changes in energy expenditure and give a high short-term precision of measurement.

2.2.3.4 Doubly-labelled water

The principle of this technique is to measure bicarbonate turnover (an index of carbon dioxide production and hence aerobic metabolism). Water labelled with deuterium (^2H) and oxygen (^{18}O) is given after measurement of a baseline urine sample. The ^2H labels the body's water pool, and the ^{18}O the combined water and bicarbonate pools. The isotopes are non-radioactive and are therefore entirely safe, even in children and pregnant women. The relative rate of disappearance of the two labels (measured by mass spectrometry) can be used to calculate the carbon dioxide production rate, and thus energy expenditure, from standard indirect calorimetry calculations (Figure 2.5).

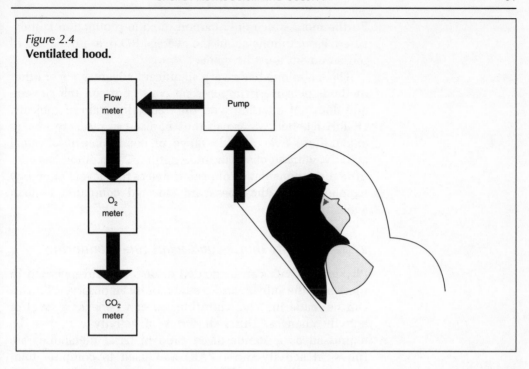

Figure 2.4
Ventilated hood.

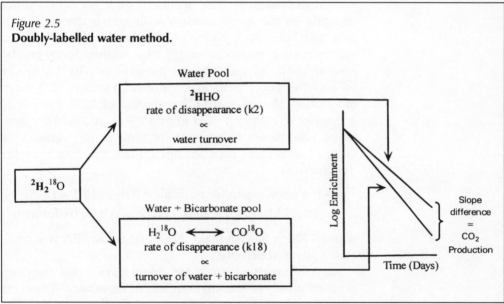

Figure 2.5
Doubly-labelled water method.

A number of assumptions and corrections have to be made with this technique: the heavier isotope may be preferentially retained compared to the naturally occurring molecules when evaporation occurs (e.g. across the lung epithelium), and some isotope may become incorporated into body solids, especially during weight gain or loss.

Furthermore, since only carbon dioxide production is measured, assumptions about the average RQ over the period of measurement must be made.

The technique has a very significant advantage over other methods because it imposes no constraints on the subject, and does not interfere with the free-living state of subjects. Doubly-labelled water can thus be used to measure energy expenditure over a wide range of conditions of physical activity, without observer intervention. The method has confirmed previous data from calorimeters that levels of energy expenditure in the obese are elevated compared to lean subjects.

2.2.3.5 *Activity diaries and heart rate monitoring*

Physical activity can be graded or detailed and recorded in diaries by the subject under study, or by an observer. Entries can be made in, say, 5 min blocks, or whenever a level of activity changes. Different forms of activity can then be expressed as a standardised ratio of basal metabolic rate (physical activity ratio, PAR) and used to compute total energy expenditure. The accuracy of these observations depends on the assiduousness with which the activity is recorded, but a further limitation is that the PAR is only an approximation to actual energy expenditure. More precise measurements of spontaneous physical activity (SPA) can be made of subjects in a respiratory chamber, with radar monitoring of movement. The amount of time in SPA in one series of studies varied from 3.7% to 19.3%. This technique was shown to account for 90% of the variance in energy expenditure between individuals, yielding a predictive formula:

$$24 \text{ h energy expenditure (kcal)} = 618 + (18.1 \times \text{FFM})$$
$$+ (10 \times \text{FM}) - (1.4 \times \text{age}) + (17 \times \text{SPA})(+ 204 \text{ for men})$$

where FFM is fat-free mass, FM is fat mass, SPA is spontaneous physical activity.

Activity monitors use a pendulum device that responds proportionately to the vigorousness of movement. These are usually worn at the waist and require an allowance for activity and exercise that disproportionately affects the upper body. The devices are small, however, and easy to wear, and give a fairly accurate measure of day-to-day variations in activity and semiquantitative data on energy expenditure.

Heart rate monitoring is based on the principle that heart rate varies with increasing oxygen consumption. For any individual it is possible to 'calibrate' oxygen consumption over a range of exercise-induced heart rates (using an alternative method such as indirect calorimetry), and then use heart rate measurements recorded from an electrocardiogram linked to a recording device. This technique is relatively cheap and can be applied in a wide variety of free-living situations.

A comparison of the varous techniques for measuring energy expenditure is given in Table 2.3.

2.2.3.6 Estimates of energy expenditure

A number of equations and formulae have been developed that predict energy expenditure from anthropometric measurements. At the most simple level RMR (resting metabolic rate) can be predicted with confidence limits of about 15% from body weight alone. Many such analyses based on regression analysis of subjects over a wide range of height, weight, and age are presented either as tables or nomograms. Formulae from a series widely used in the UK are given in Table 2.4.

Total 24 h energy expenditure can be approximated from estimates or measures of resting energy expenditure according to the following:

- 1.3 × RMR (inactive, sedentary lifestyle with no sport or active leisure pursuits).
- 1.5 × RMR (average British person, sedentary occupation, little strenuous activity at home, occasional sport).
- 1.7 × RMR (more active than average due to manual occupation or regular strenuous leisure pursuits or significant time spent walking or cycling).

2.2.3.7 Estimates from weight change

In clinical settings energy intake and body weight change can be used as a crude estimate of energy expenditure. The composition of weight slowly gained or lost is about 75% fat and 25% fat-free tissue. It has been found experimentally that the metabolisable energy of fat is 9 $kcal\,g^{-1}$, monosaccharides 3.75 $kcal\,g^{-1}$ and starch 4 $kcal\,g^{-1}$ (close to the values predicted from their chemical formulae). The gross energy of protein is 5.25 $kcal\,g^{-1}$, but metabolisable energy accounting for urea lost in the urine is nearer 4 $kcal\,g^{-1}$. From these factors, it can be seen that 1 kg of weight lost provides about 7700 kcal. Thus a subject on a daily 1000 kcal diet who loses 1 kg weekly must be using approximately

Table 2.3 Advantages and disadvantages of different techniques for measuring energy expenditure

	Direct calorimetry	Indirect calorimetry			Doubly-labelled water	Heart rate, activity diaries
		Chamber	Douglas bag	Ventilated hood		
Accuracy	<1%	1–2%	3%	2%	4%	5%
Response time	Fast	Fast	Moderate	Very fast	None	Moderate
Information on substrate metabolism	None	Yes	Yes	Yes	None	None
Constraints on activity	Slight	Slight	Moderate	Moderate	None	Slight
Duration of measurements	Days to weeks	Days to weeks	Hours	Hours	2–3 weeks	Days to indefinite
Expense	Very high	Very high	Moderate	Moderate	Very high	Cheap

Table 2.4 James and Lean formulae for calculating resting metabolic rate in kcals per 24 h

Age (years)	Female	Male
>18 and <30	$487 + (14.8 \times weight)$	$692 + (15.1 \times weight)$
>30 and <60	$845 + (8.17 \times weight)$	$873 + (11.6 \times weight)$
>60	$658 + (9.01 \times weight)$	$588 + (11.7 \times weight)$

2100 kcal daily (1000 kcal from dietary intake and 1100 kcal from weight loss). These calculations are easy to make at the 'bedside' and in reverse can be a useful approximation for the calorie deficit a subject has achieved.

2.3 Altered energy expenditure in obesity

Although there is agreement that total energy expenditure is increased in established obesity, evidence from animal models of obesity, and some from humans, suggests that components of energy expenditure may be altered and impaired in the preobese state.

2.3.1 *Animal models*

A large number of genetically determined forms of obesity have been identified, especially in rodents. These models vary in their mode of inheritance, the type of obesity, as well as the metabolic and physiological associates (Table 2.5). Single-gene, polygenic and dietary/polygenic interacting models of obesity in mice and rats are well characterised and can be used to investigate the physiological mechanisms involved in the development of obesity. With advances in molecular biology over the past few years, the genes themselves can be identified and make it possible to identify the gene product and hence the physiological effector of the obesity.

Table 2.5 Single gene mouse models of obesity

Mouse locus		Transmission	Chromosome	Human homologous region	Status
Diabetes	(*db*)	Recessive	4	1p31	OB receptor cloned
Obese	(*ob*)	Recessive	6	7q31	Cloned, codes for leptin
Tubby	(*tub*)	Recessive	7	11p15	Cloned
Fat	(*fat*)	Recessive	8	16q22	Cloned
Agouti	(*A^y*)	Dominant	2	20q13	Cloned

A number of features are common to these animal models. In many such examples, obesity develops at birth together with hyperphagia and insulin hypersecretion. Excessive food intake has been shown not to be essential for the development of obesity. Since physical activity is normal, at least until the final stages of obesity, it can be inferred that these animals must have some reduction in a component of energy expenditure. In some models, resting metabolic rate is reduced, while in others brown adipose tissue thermogenesis (important in temperature regulation of rodents) is impaired or absent. For example, lean litter mates of *ob/ob* mice fed a highly palatable diet increase energy consumption by 69% but do not increase their rate of weight gain. In contrast, the *ob/ob* littermates in this experiment overconsumed by 50% and retained more than half of the excess energy as body fat. The link between hyperphagia and energy expenditure suggests that the hypothalamus may be involved in regulating both components of energy balance, and there is much experimental evidence to support this hypothesis. The hypothalamic peptide neuropeptide Y (NPY) and corticotrophin-releasing hormone appear to be important in this integrated control.

A large body of work had already pinpointed the hypothalamus in body weight regulation, and that its influences on feeding and energy expenditure are closely linked. In very brief summary, lesions of the ventromedial hypothalamus induce both hyperphagia and reduce meal-induced thermogenesis. The methods of lesioning and the precise location of the hypothalamic damage determines the nature of the physiological alterations (diet, hyperphagia, thermogenesis, insulin resistance). The variability of findings results from the particular damage caused to vagal and sympathetic pathways bringing peripheral signals to the hypothalamus and relaying efferent information to the pancreas and brown adipose tissue. Damage to neural pathways explains the findings better than older concepts of specific 'feeding centres' or 'satiety centres'.

The cloning of the *ob* gene led to its product, leptin, being identified. This new hormone is present in blood in quantities that are closely linked to fat cell mass, and is believed to signal the hypothalamus about the nutritional state of the organism. In *ob/ob* mice, the *ob* gene is mutated and fails to produce leptin; these mice are hyperphagic and become obese. Administration of leptin to *ob/ob* mice reverses the hyperphagia and can produce weight loss. It seems clear now that the *db/db* mice have mutations in the leptin receptor, since leptin levels are high but ineffective at regulating feeding behaviour. The link between leptin as a satiety hormone and the hypothalamic control of food intake is not proven yet. This research appears to point to feeding behaviour rather than energy expenditure as being the primary controller of body weight. However, as mentioned earlier, the hypothalamic influences on feeding also affect sympathetic nervous system (SNS) activity which is in turn an important regulator of energy expenditure. Increased SNS activity has been shown in normal animals to increase thermogenesis in response to cold and overfeeding, whilst in genetically obese animals SNS activity is reduced.

A significant limitation of these animal models of obesity is that they relate poorly to the more common adult-onset obesity in man. Furthermore, the multiplicity of proposed mechanisms to account for observations across the animal models of obesity remains confusing, and it is often unclear which events are the primary cause for weight gain. There are, however, increasing data in man that parallel the animal findings.

2.3.2 *Human obesity*

2.3.2.1 *Energy balance*

Energy balance is achieved with remarkable accuracy, even in the obese. An average-sized man gaining 7 kg over 10 years will have matched about 10 million kcal of energy intake with expenditure to an accuracy of more than 99.5%. The relationship of energy balance to energy stores in a static state can be expressed as:

Energy intake − Energy expenditure

= Change in energy stores

This would predict that a small daily increase in energy intake of, for example, 100 kcal would lead to a weight gain of 189 kg over a 40-year period. That this does not occur could suggest that there are energy-wasting mechanisms that guard against weight gain, but a simpler explanation is that the equation does not relate to situations of energy imbalance. Since weight gain leads to increased energy stores and body weight, which themselves consequently increase energy expenditure, it is more useful to think of energy balance as a dynamic situation expressed as:

Rate of energy intake − Rate of energy expenditure

= Rate of change in energy stores

From this equation it is seen that a positive energy balance will increase fat-free mass, which will in turn increase energy expenditure to lead to a new balance of body weight at a stable, but higher level. A further prediction from this hypothesis is that chronic states of imbalance, rather than initial defects in energy intake or expenditure, are more likely to be found in, and account for, obesity. However, until recently, most work has concentrated on initial or underlying defects.

A number of epidemiological studies have suggested a lower energy intake in obese compared to lean individuals. However, this evidence is flawed by its reliance on self-reported food intake—data which are notoriously unreliable. The advent of the doubly-labelled water technique allowed the accuracy of energy intake measures to be directly assessed under free-living conditions, and has shown reported intakes to under- or overestimate intakes by 50% (Figure 2.6).

Many other studies (e.g. in calorimeters) have confirmed that energy expenditure is higher both at rest and in response to activity in established obesity. These findings do not, how-

Figure 2.6
Accuracy of reporting dietary intake from 7-day weighed intake records using doubly-labelled water as index of energy expenditure (data from Livingstone, M.B.E., Prentice, A.M., Coward, W.A., Ceesay, S.M., Strain, J.J., McKenna, P.G., Nevin, P.G., Barker, M.E., and Hickey, R.J., 1990. Simultaneous measurement of free-living energy expenditure by the doubly-labelled water ($2H_2^{18}O$) method and heart rate monitoring. *American Journal of Clinical Nutrition*, 52, 59–65.

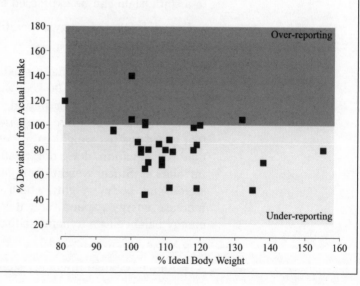

ever, rule out the possibility that defects in energy expenditure could contribute to, or be responsible for, the development of some human obesities. A number of prospective studies have suggested that a lower than average energy expenditure predicts future weight gain in children aged 3–5 years, in the offspring of lean and obese mothers and in subgroups such as the Arizona Pima Indians. However, since individuals with the lowest absolute energy expenditure are the thinnest (by virtue of their low body weight), one must invoke a concomitant failure of downregulation of food intake as being necessary for the development of obesity in those with relatively low energy expenditure.

2.3.2.2 Physical activity and thermogenesis

Cross-sectional population-based evidence suggests that, in the UK, average energy expenditure from physical activity has declined faster than average energy intake. This is strong evidence for a permissive role of inactivity in the pathogenesis of obesity. In established obesity, there is little evidence to show reduced levels of spontaneous activity, and at every level of activity the obese have a higher energy expenditure than the lean. Thus there is little to support the contention that the obese remain obese because of inactivity.

There is conflicting evidence over thermogenesis in obesity. One of the most convincing studies compared thermic responses to low- and high-calorie mixed meals and to ephedrine, a pharmacological stimulus, in lean (BMI (body mass index) 21.5) and modestly obese (BMI 26) young men. Thermic responses to all stimuli were lower in the obese. A similarly reduced thermic response to food or sympathetic stimulation has been demonstrated after rapid, but not gentle, weight loss, or after repeated bouts of weight loss.

While the above are all of physiological interest, it seems hard to believe that these defects have much to do with human obesity development. The degree of hypometabolism is very small, affects a very small component of total energy expenditure, and would in general be outweighed by the much greater resting energy expenditure that is determined by the higher mass (and fat-free mass) in the obese.

The mechanisms of reduced thermogenesis are of considerable interest in that they relate both to the insulin resistance seen in obesity and to reduced to SNS activity. During a hyperinsulinaemic, euglycaemic clamp, reduced thermogenesis and net glucose storage can be seen in obese individuals (with normal glucose tolerance). Normalising their rate of glucose storage, by further raising insulin levels, normalised the thermic response. Reduced SNS activity preceding the development of obesity is prevalent in some racial groups (e.g. Pima Indians), and can be demonstrated under a variety of conditions. The lower thermogenesis seen in some studies after weight reduction could be best explained by reduced SNS activity rather than by insulin resistance, which would be expected to improve after weight loss. To add to evidence that energy intake and expenditure are tightly linked, there is evidence that meal-induced thermogenesis enhances satiety in man.

2.3.2.3 *Nutrient influences*

Farmers have long known that a high-fat diet is a potent way of increasing animal carcass weight. There has been a large increase in dietary fat consumption this century (from 20% to 40% of total energy), with a concomitant decrease in energy from carbohydrate, and there has been a steady rise in the prevalence of obesity. One suggestion is that dietary fat may have a less satiating effect on a per calorie basis than carbohydrate, thus encouraging a higher calorie intake; but energetic considerations also suggest that a high fat diet may directly encourage weight gain.

The energy balance equations (p. 37) are a simplification since it would seem self-evident that intake and oxidation rates for each macronutrient must also balance. There is a hierarchy of storage capacity in the body, from nil for alcohol to unlimited for fat (Table 2.6).

Since alcohol must be obligatorily metabolised it cannot directly cause obesity, except by contributing to overall energy balance and sparing some other component (i.e. fat oxidation). A similar situation exists for protein balance, since a positive balance must imply increased muscle (or fat-free mass) and this only occurs in response to physical training, growth or secondary to weight gain. In most societies carbohydrate is the largest source of dietary energy. Stores of carbohydrate (mainly as liver glycogen) are limited

Table 2.6 Storage of macronutrient energy substrates

	Alcohol	Carbohydrate	Protein	Fat
Daily energy intake	0–5%	30–50%	15–20%	30–50%
Size of stores	None	Tiny	Moderate	Unlimited
% Intake/store	0%	50%	1%	<1%
Stores regulated		Yes	Yes	No
Potential for imbalance	None	Slight	Only with growth	Yes

and can be depleted rapidly over days. Tight control of glucose oxidation and glycogen storage would be expected, since only in extreme conditions can positive carbohydrate energy balance occur and produce de novo lipogenesis. If this is so, then mechanisms might also exist to allow feeding behaviour to sense carbohydrate balance and alter macronutrient choice accordingly. A number of signals of carbohydrate balance have been proposed; insulin is the prime candidate.

In the case of fat, where stores are essentially unlimited, one might expect weak regulation of fat balance and little metabolic response to a positive balance. This is indeed the case. Only small increases in fat oxidation rates occur with fat overfeeding. What does increase fat oxidation is a negative energy balance, to such a degree that fat balance can be regarded as synonymous with total energy balance. In this respect, obesity could be seen as an adaptation to chronic overfeeding (especially with a high-fat diet). While thermodynamic principles must be maintained, overfeeding with fat especially disposes to weight gain and accounts for the finding of high fat intakes amongst obese individuals and in communities with a high incidence of obesity.

2.4 Pharmacological approaches to increasing energy expenditure

Conventional dietary treatment of obesity is unsatisfactory, both in terms of success at initial weight reduction and, more particularly, in terms of weight loss maintenance. Increasing exercise and activity are valuable components of any weight control programme (and have been shown to correlate strongly with success at weight loss maintenance), but it is often difficult for the severely obese patient to be physically active. Such a patient also has as much of a motivational problem about changing this area of their behaviour as over voluntarily changing eating habits. Viewed as a chronic disease that causes morbidity and mortality, there is no reason why pharmacological approaches to treatment should not be considered. At present, most drugs used to treat obesity are anorectic agents and work mainly to decrease food intake. They are usually given in conjunction with advice on a reduced energy diet. The effect of dietary restriction, whether voluntary or drug-induced, must inevitably wane with time as weight is lost and metabolic rate falls. Most existing drugs have been shown in randomised trials to produce a mean weight loss of about 10 kg, with a plateau of weight being reached after about 6 months' of treatment.

For the past 20 years, there has been a concerted search for drug treatments that might safely increase metabolic rate in obese patients. An important consideration, however, is their potential effects on heart rate. Many obese patients have already, or are at risk from, ischaemic heart disease and hypertension; thermogenic drugs would have to minimise any effect that such hypermetabolism could have on increasing cardiac output and myocardial oxygen consumption, and on blood pressure.

The first synthetic thermogenic drug to treat obesity was dinitrophenol, which by uncoupling oxidative phosphorylation from ATP formation lead to substrate oxidation with the energy being 'wasted' as heat. Unfortunately, during the 1930s when it was in use, there were a number of deaths and it fell into disrepute.

2.4.1 *Thyroid hormones*

Thyroid hormones (thyroxine and tri-iodothyronine) have been known since the turn of the century to increase metabolic rate and thermogenesis. Patients with spontaneous overactivity of thyroid hormone production (thyrotoxicosis) have increased resting metabolic rate, often lose weight (in the face of increased appetite and food intake), and report increased sensations of heat and sweating. The mechanisms by which thyroid hormones (and in particular tri-iodothyronine) increase metabolic rate are not fully understood but involve binding to nuclear receptors. The conundrum concerning their action is that thyroid hormone-induced increases in respiration are tightly coupled to the utilisation of high-energy phosphorus compounds (e.g. ATP), but it has been hard to identify the biochemical processes that use the ATP and produce enough ADP to feed the high oxidation rates. Different organs may respond in different ways to thyroid hormone action, but enhanced Na^+, K^+-ATPase pumping across cell membranes, increased mitochondrial oxidative phosphorylation, hepatic lipogenesis and sensitisation of brown adipose tissue to SNS stimulation have all been described.

In clinical practice, thyroxine and tri-iodothyronine are not useful for treatment. While they increase metabolic rate, they cause tachycardia and may provoke dysrhythmias or myocardial infarction, and are also associated with accelerated protein (fat-free mass) loss. Research into thyroid hormone analogues that avoid these unwanted effects is in progress.

2.4.2 *Ephedrine/caffeine*

Phenylpropanolamines such as ephedrine, and xanthine derivatives such as caffeine, increase metabolic rate. The amount of caffeine contained in two cups of coffee (100 mg), will raise the metabolic rate by about 4%, but the effect is short-lived, and may be followed by a period of reduced energy expenditure. Both ephedrine and caffeine appear to act synergistically, and this combination has been marketed in Denmark as the Elsinore pill. The actions of ephedrine are mediated through sympathetically released noradrenaline, which could act at β_3-receptors (on brown adipose tissue), β_2-receptors (stimulating protein synthesis and increasing lean body mass) and at postsynaptic α-receptors involved with conversion of thyroxine to triiodothyronine. Xanthines (and aspirin) inhibit negative feedback of released noradrenaline, adenosine and prostaglandins on noradrenaline release from sympathetic nerve terminals. Clinical trials have shown these drugs alone, or combination, to be effective at producing weight loss in obese subjects. There use is limited by the finding that the drugs also raise blood pressure and increase heart rate.

2.4.3 *Atypical β-receptor agonists*

The finding of atypical (β_3) receptors that mediate the thermogenic effects of sympathomimetic agents, but not the β_1 effects of heart rate stimulation, or β_2 effects on smooth muscle contraction and tremor, suggested a potentially valuable approach for thermogenic drug development. These receptors are found predominantly on brown adipocytes and have now been cloned. A number of compounds have been shown to be effective in stimulating energy expenditure and heat production from brown fat in rodents, but in man the results have been disappointing. The drugs have had little effect and have produced unwanted effects such as muscle tremor. It is known that the human β_3-receptor differs from the rodent receptor, and new more-specific compounds are being developed. However, there is little evidence that significant numbers of brown adipocytes exist in adult humans. The assumption must be that activation and hypertrophy or proliferation of these cells may be induced by chronic specific β_3 stimulation if the drugs are to be clinically useful.

2.5 Conclusions

Obesity is at one end of a spectrum of body weight, from malnourishment with inadequate energy stores, through 'ideal' body weight, to a state of excessive stores associated with health impairment. Obesity is increasing in our society because human physiology is ill-adapted to changes that have favoured increased energy consumption and decreased energy expenditure. An understanding of the regulation of body weight and energy fluxes is needed if the precise predisposition and causes of excessive fat storage are to be elucidated, and effective treatments developed. At a practical level, it is important for doctors, dietitians and their patients to realise that weight loss can only be achieved by a negative energy balance, and that low levels of energy expenditure do not constitute a bar to achieving this therapeutic goal.

Further reading

Garrow, J.S., 1988. *Obesity and Related Diseases*. Edinburgh: Churchill Livingstone.

Murgatroyd, P.A., Shetty, P.S. and Prentice, A.M., 1993. Techniques for the measurement of human energy expenditure: a practical guide. *International Journal of Obesity*, **17**, 549–568.

Swinburn, B.A. and Revussin, E., 1994. Energy and macronutrient metabolism. *Baillière's Clinics in Endocrinology and Metabolism*, **8**, 527–548.

3 Diabetes Mellitus and its Complications

Michael E. Edmonds

3.1 Introduction

The manifestations of diabetes result from a persistently raised blood glucose level as a consequence of reduced production and/or impaired effectiveness of insulin. Diabetes can be divided into two main groups: insulin-dependent diabetes (IDDM), and non-insulin-dependent diabetes (NIDDM). In IDDM, there is an almost complete lack of effective insulin and, in the absence of insulin treatment, these patients will usually progress to diabetic ketoacidosis. In NIDDM, there is a relative, but not an absolute, lack of insulin. Peripheral tissue becomes insulin-resistant (i.e. less sensitive to the effects of insulin).

This chapter will consider the molecular aspects of diabetes and is divided into four sections:

1. Insulin biosynthesis and function, and insulin analogues.
2. Molecular genetics and the aetiology of insulin-dependent diabetes.
3. Molecular genetics and the aetiology of non-insulin-dependent diabetes.
4. The molecular basis of diabetic complications.

3.2 Insulin biosynthesis and function

3.2.1 The beta cell

The beta cell is the major endocrine cell component of the islets of Langerhans, and a typical human islet will contain 3000–4000 cells. The relative proportions of the different endocrine cell types may vary in different anatomical areas of the pancreas, but on average there are 70% beta cells, 25% alpha cells and 5% D cells. The beta cell is programmed to synthesise insulin via its precursor pro-insulin, and furthermore, it has developed detailed mechanisms to regulate the rate of secretion in response to glucose. A characteristic feature of a beta cell is the presence of a large number (up to 13 000 per cell) of membrane-limited hormone-storage granules or vesicles which contain a substantial reservoir of insulin, sufficient for 24 h of secretion in the absence of new synthesis. The dense core of the storage granule contains insulin in a crystallised form, along with the connecting peptide excised during the process of conversion from the biosynthetic precursor pro-insulin. These two together make up 80–90% of the total granular protein.

The beta cells also synthesise a number of other proteins made in much smaller quantities than insulin, including a peptide called islet amyloid polypeptide (IAPP) or amylin. This is exported from beta cells along with insulin, in insulin granules, and can be deposited outside the cells as amyloid fibrils. It is not yet known whether these deposits, which can be found around beta cells in some patients with NIDDM, are a cause or effect of hyperglycaemia.

3.2.2 *Insulin biosynthesis*

The insulin molecule consists of two polypeptide chains— the A chain (21 amino acid residues) and the B chain (30 residues)—linked by two disulphide bridges. Human insulin differs from porcine insulin in a single residue (at B30) and from bovine insulin by two additional amino acid substitutions (at A8 and A10).

The process via which insulin is synthesised in the beta cell of the pancreatic islet, so that the peptide can be secreted as a correctly folded, active hormone upon appropriate beta cell stimulation, is critical to insulin function. The stages of insulin biosynthesis are shown in Figure 3.1.

The human insulin gene is found on the short arm of chromosome 11. It corresponds to about 150 base pairs (bp) in length and contains the following sections:

- A 5′ untranslated region that is split by an intervening sequence.
- A region corresponding to the sequence of a signal peptide that functions in the translocation of the hormone precursor across the membrane of the endoplasmic reticulum so that the peptide enters the secretory pathway.
- A region corresponding to the insulin beta chain.
- A region corresponding to the C-peptide which is again split by an intervening sequence.
- A region corresponding to the insulin A chain.
- A 3′ untranslated region.

Figure 3.1
The structure of the insulin gene and the production of insulin through gene expression and precursor processing. U, untranslated region; I, intervening sequence; S, signal peptide; B, insulin B chain; C, pro-insulin connecting peptide; A, insulin A chain.

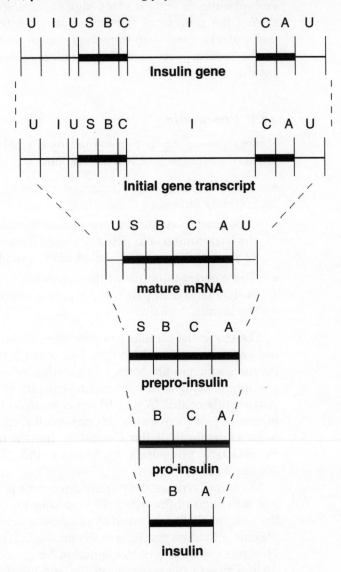

Transcription of the insulin gene leads to a complementary mRNA precursor of the same size which is then modified by capping of the 5′ end by 7-methylguanosine, by excision of the intervening sequences and by polyadenylation at the 3′ end. The mature mRNA then leaves the nucleus and is translated on membrane-bound ribosomes of the rough endoplasmic reticulum. The initial amino-terminal segment of the protein (the so-called signal peptide) helps to guide the translocation of the corresponding protein, prepro-insulin, across the membrane into the lumen of the endoplasmic reticulum. This signal sequence is removed during the process of translation to yield the pro-insulin molecule. Glucose is an important regulator of transcription of the insulin gene and of translation of the mature pro-insulin mRNA.

3.2.3 *Pro-insulin*

Human pro-insulin is represented by a single polypeptide chain of 86 amino acid residues (Figure 3.2) in which:

- Disulphide bonds involving cysteine residues have been correctly formed.

- The carboxy-terminus of the insulin B chain is linked by the basic amino acid pair Arg-Arg to the amino-terminus of a connecting peptide called the C-peptide.

- The carboxy-terminus of the C-peptide is linked by the basic amino acid pair Lys-Arg to the amino-terminus of the insulin A chain.

The molecular structure of pro-insulin contains within it the entire structure of insulin. The major function of pro-insulin is to permit correct orientation of the interchain disulphide bridges in the insulin structure so that the two polypeptide chains (A and B) are completed correctly. The processing or conversion of pro-insulin to insulin then requires removal of the C-peptide domain of pro-insulin by selective proteolysis to generate the hormone itself (Figure 3.2).

The initial synthesis of pro-insulin occurs in the endoplasmic reticulum of the beta cell. Prepro-insulin derived from the mature mRNA is cleaved by protease activity in the endoplasmic reticulum resulting in the production of pro-insulin. This process is subject to regulation by glucose and cyclic AMP. Once in the cisternae of the endoplasmic reticulum, pro-insulin rapidly achieves its three-dimensional configuration, with the disulphide bridges correctly formed. It is then transported via microvesicles to the Golgi complex where concentration of the pro-insulin in membrane-limited sacs is initiated.

As the peptide hormone precursor moves to and through the multiple compartments of the Golgi apparatus, it is mixed with endo- and exoproteinases which are necessary

Figure 3.2
Conversion of pro-insulin to insulin through enzyme-catalysed precursor processing. The pairs of basic amino acids that link the insulin beta chain to the C-peptide (Arg-Arg) and C-peptide to the insulin A chain (Lsy-Arg) are shown.

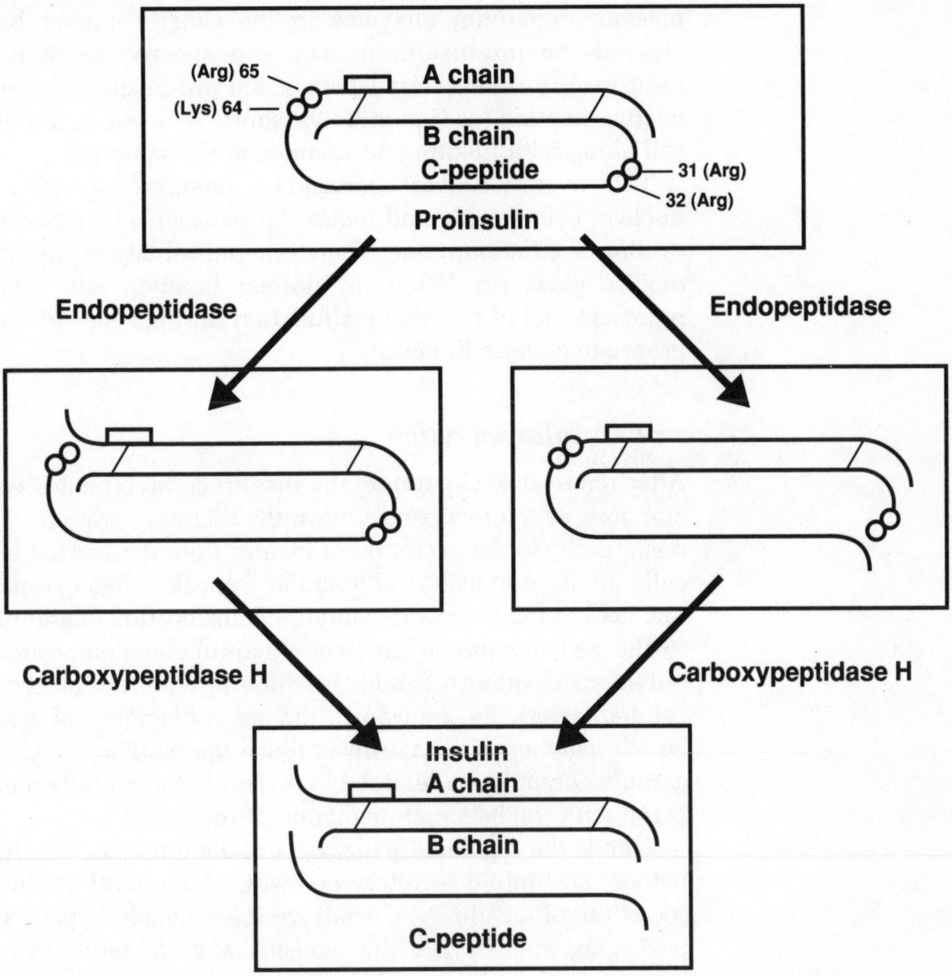

for conversion of the precursor to insulin. The secretion granules contain three enzymes responsible for the conversion of pro-insulin to insulin. They comprise two categories: endopeptidases and carboxypeptidase H. Endopeptidases catalyse cleavage at the Lys-Arg pair and the Arg-Arg pair of basic amino acids to yield pro-insulin processing intermediates. Following proteolytic cleavage of pro-insulin by endopeptidases, the basic amino acids at the junctions must be removed. This is accomplished by carboxypeptidase H, thereby achieving the final insulin structure.

Conversion of pro-insulin to insulin begins at the distal portion of the Golgi apparatus and then continues to near completion in secretion granules that are pinched from the Golgi membrane to yield insulin plus the pro-insulin C-peptide. Within 30 min of its synthesis in the beta cell endoplasmic reticulum, enzymes in the Golgi complex have cleaved the pro-insulin molecule at specific residues to form insulin. The excised part of the pro-insulin, the connecting peptide (or C-peptide for short) is stored in the beta cell along with insulin and secreted at the same time.

This overall process demands a complex interplay of nuclear, cytoplasmic and molecular mechanisms in the biosynthesis of the hormone. There is a multiplicity of sites that require close regulation for normal function and which represent loci of possible malfunction during states of inappropriate glucose homeostasis.

3.2.4 *Insulin secretion*

After removal of C-peptide, the insulin co-precipitates with zinc ions as microcrystals within the secretory granule. The major pathway for secretion of insulin from the normal beta cells involves transport of granules from the large cytoplasmic pool to the plasma membrane. Translocation of granules to the cell membrane involves microtubules composed of polymerised tubulin subunits, which provide the mechanical framework for transport, and microfilaments of actin, which together with myosin produce the motive force. The granule contents are released by fusion of the granule membrane with the beta cell membrane (exocytosis).

This is the regulated pathway, but there may be an alternative constitutive secretory pathway which involves direct secretion of insulin via small vesicles which bypass the packaging and storage mechanisms and therefore bypass the conversion process. This is probably not an important pathway in normal beta cells, but could be relevant in early IDDM and NIDDM in which high levels of circulating pro-insulin have been detected.

3.2.5 *Regulation of insulin secretion*

The major physiological determinant of insulin secretion is glucose availability, although physiological and pharmacological agents act as secretagogues. These can be divided into two groups: initiators or primary stimuli (i.e. agents capable of provoking insulin release alone) and potentiators, which are ineffective alone, but which will increase the insulin released in response to glucose or hormones such as glucagon.

The most important factor in the regulation of insulin secretion is change in blood glucose concentrations, and beta cells have a very sensitive mechanism which allows them not only to secrete insulin when the blood glucose levels are above fasting but to increase rates of secretion very considerably as blood glucose concentrations increase in the range of 6–12 mmol l^{-1}. These responses are very sensitive and rapid and occur usually within 30 s of an initial glucose stimulus. Glucose-stimulated insulin release is biphasic, comprising a rapid first phase lasting 5–10 min and a prolonged second phase lasting throughout the duration of the stimulus. The characteristic response to glucose is achieved largely by the enzyme glucokinase, which is an important regulating step in the early stages of beta cell glucose metabolism. Genetic defects in some of the enzymes of glucose metabolism may lead to defective patterns of insulin secretion in NIDDM, and this mechanism has been identified in the rare subset of dominantly inherited NIDDM or maturity-onset diabetes in the young (MODY) (see below).

Gut hormones, including gastric inhibitory polypeptide, can act to potentiate insulin secretory responses to glucose by raising concentrations of cyclic AMP within the beta cell. This sensitizing effect on the beta cell of gut hormones released in the process of digestion could explain the increased insulin secretory response found to oral glucose compared to intravenous glucose. Neurotransmitters are also potentiators of insulin secretion. The pancreatic islets are innervated by the autonomic nervous system and stimulation of the vagus entering the pancreas causes insulin release which can be blocked by atropine. Furthermore, insulin secretion can also be evoked by direct stimulation with acetylcholine, and this can also be blocked by atropine. These secretory effects are dependent on the presence of glucose and calcium uptake. In contrast, insulin secretion is inhibited following stimulation of sympathetic innervation and also by adrenaline and noradrenaline.

3.2.6 *Insulin receptor*

The interaction of insulin with its receptor triggers the hormone's effects on intracellular enzymes from the cell surface. The intact insulin receptor consists of a heterodimer of four glycosylated peptide subunits covalently linked by disulphide bonds. Two identical alpha-subunits (which contain the insulin binding sites) are linked covalently to two beta-subunits which straddle the cell membrane (Figure 3.3).

The gene for the insulin receptor precursor lies on the short arm of chromosome 19. Transcription yields several mRNA species varying in length from 5.7 to 9.5 kb. Translation on the ribosomes leads to the pro-receptor peptide. At the amino-terminal, a signal sequence of 27 hydrophobic amino acids precedes the alpha-subunit; then follows a tetrapeptide typical of proteolytic cleavage sites and, finally, the 620 amino acids of the beta-subunit. Both subunits contain glycosylation sites related to asparagine residues. After glycosylation and cleavage, the receptor is inserted into the plasma membrane.

Binding of insulin to the alpha-subunit is accompanied by autophosphorylation of tyrosine residues of the beta-subunit and subsequent activation of a protein tyrosine kinase in the beta-subunit. This in turn triggers changes in insulin receptor substrates which cause activation of the enzymes that catalyse the various metabolic effects of insulin.

3.2.7 *Insulin resistance*

Insulin resistance can be defined as a state in which a normal amount of insulin produces a subnormal biological response. Reduced insulin effect may be due to changes in the insulin molecule, circulating antagonists or target cell defects. Thus, there are three general causes of insulin resistance:

1. Abnormalities of beta cell secretory products.
2. Prereceptor antagonists of insulin action.
3. Target cell defects at the receptor or postreceptor level.

3.2.7.1. *Abnormalities of beta cell secretory products*

Abnormal beta cell secretory products include abnormal insulin molecules (or the mutant insulin syndrome) and hyperproinsulinaemia. In the mutant insulin syndrome, there is hyperinsulinaemia and reduced biological activity of endogenous insulin. Hyperproinsulinaemia is characterised by high plasma insulin values, due to cross-reactivity with pro-insulin, but impaired glucose intolerance. The cause of the disease is a reduced conversion of pro-insulin into C-peptide and insulin.

Figure 3.3

(a) Insulin receptor gene. (b) Insulin receptor protein. The insulin receptor protein is made up of four polypeptide chains, two α chains which are extracelluar and two β chains which span the cell membrane. The COOH terminals of the α chains are covalently bound to the NH₂ of the β chains.

(a) Insulin receptor gene

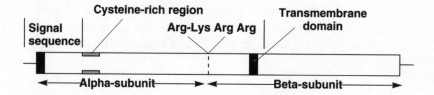

(b) Insulin receptor protein

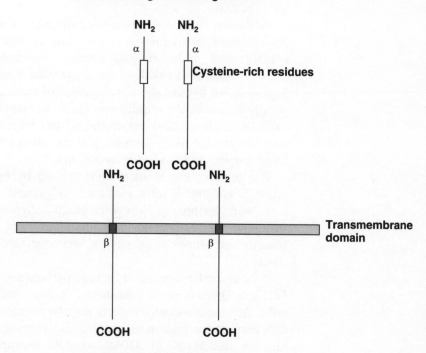

3.2.7.2 Circulatory or prereceptor antagonists of insulin action

Prereceptor antagonists of insulin action include circulating insulin antagonists of two types, immunological and hormonal. Immunological antagonists include autoantibodies to insulin, which limit the access of insulin to its receptor and thereby induce insulin resistance, and autoantibodies to the insulin receptor, which also lead to insulin resistance. Hormonal antagonists include increased concentrations of cortisol, growth hormone, glucagon, catecholamines, thyroid hormone, prolactin and androgens.

3.2.7.3 Target cell defects

It is now generally accepted that the cause of insulin resistance in most patients is some defect in cellular insulin action either at the receptor or at postreceptor level. Several conditions are characterised by cellular insulin resistance; these include the insulin resistance syndrome, obesity, NIDDM.

Patients with NIDDM are often hypertensive and dyslipoproteinaemic, with increased very low-density lipoprotein (VLDL) and reduced high-density lipoprotein (HDL). Insulin resistance in skeletal muscle may be a primary aetiological factor behind the development of all these phenotypically characteristic conditions. Thus, the insulin resistance syndrome, also called syndrome X, has been formulated, comprising insulin resistance, glucose intolerance, arterial hypertension and dyslipoproteinaemia.

The most common state characterised by insulin resistance is obesity. Insulin resistance is present in both the liver and periphery. However, chronic hypocaloric dieting may normalise the insulin effect, suggesting that insulin resistance in obesity is secondary to the obesity itself.

Insulin resistance is a major pathogenetic factor in NIDDM. There is insulin resistance in liver, muscle and fat cells. Although receptor defects may be responsible for certain parts of the insulin resistance, postreceptor defects may also be important. In IDDM, absolute insulin deficiency represents the primary pathogenetic abnormality. However, most IDDM patients are also moderately or severely resistant to the actions of insulin both at the onset and after long-standing duration of the disorder.

3.2.8 *Synthesis of human insulin and insulin analogues*

Human insulin was manufactured in the mid-1980s by recombinant DNA technology in which *Escherichia coli* or yeast was programmed to produce insulin. Three strategies have been adopted. The first strategy provides for separate A and beta chain fermentation in *Escherichia coli*, the second for insertion of the entire pro-insulin gene into a single organism rather than as separate A and B chains, and the third, for biosynthesis of a single-chain precursor which contains a modified C-peptide region, thus ensuring correct folding. Whatever the method of production, human insulin has a structure identical to that of the native hormone.

Study of the structure–function relationship of insulin has also facilitated the production of insulin analogues with advantageous duration of action and differential effects on hepatic and peripheral glucose disposal. The latest development is a set of analogues of insulin prepared by peptide synthesis, enzyme-catalysed semisynthesis and recombinant DNA technology. This is important because the pharmacokinetic-dynamic effects of the presently available commercial insulin preparations are incapable of bringing about near normal glycaemia without significant hypo- as well as hyperglycaemia. The aim of insulin analogues is to simulate better the pharmacokinetics of normal insulin secretion with subcutaneously injected exogenous insulin analogues and thereby promote more efficient glucose metabolism in fed and fasted states.

Insulin's ability to self-associate is important both in analysing the structure of the hormone and in evaluating various insulin preparations for potential therapeutic value. Human insulin exists as monomers only in very dilute solutions; at higher concentration insulin occurs in solution mainly as a mixture of dimers, hexamers and even more highly associated forms. In the presence of zinc ions three insulin dimers (a total of six molecules) associate to form a two-zinc insulin hexamer. Insulin is known to occur in the secretory granules of the pancreatic islet in a zinc-coordinated microcrystalline state, and it is likely that the hexameric structure of the hormone is the structure of insulin as it exists at its site of synthesis and storage in vivo. The relative stabilities of the various hexameric and crystal forms of insulin have therapeutic implications. Stable forms of insulin hexamers will be absorbed slowly from sites of administration (the so-called slow-acting insulin preparations), whereas unstable forms of insulin hexamers (or insulin dimers or monomers) will be absorbed rapidly (the so-called fast-acting insulins).

Attention has been focused on the production of monomeric insulins that might act faster than so-called regular insulins. High-potency monomeric forms of insulin will be absorbed quickly and would allow the control of blood glucose to be modulated. Since natural insulins self-associate under any therapeutically relevant condition, the search for monomeric insulins depends on alterations in the structure and can be accomplished by replacement of one or more of insulin's 51 amino acid residues by protein engineering. A further target of simulation by insulin analogue development programmes is the low insulin concentration that occurs normally in the basal or postabsorptive state and which functions to restrain hepatic glucose production. Various modifications have led to the development of intermediate- or long-acting insulins, although progress has been slower than with fast-acting insulin analogues.

3.3 Molecular genetics and aetiology of insulin-dependent diabetes (IDDM)

3.3.1 Introduction

It was previously thought that IDDM was an acute-onset disease, presenting with polyuria, polydipsia, hyperglycaemia and ketoacidosis. However, IDDM is a chronic autoimmune disorder that gradually develops over many years. A variety of abnormalities in immune function and insulin release precede the abrupt development of the diabetic syndrome in patients genetically predisposed to diabetes. Therefore, there are several stages in the natural history, including genetic susceptibility, triggering events, active autoimmunity, gradual loss of glucose-stimulated insulin secretion, appearance of overt diabetes with some residual insulin secretion and complete beta cell destruction.

3.3.2 Molecular genetics

The exact mode of inheritance of IDDM is complex and not completely understood, partly because of the heterogeneity of diabetes. There is no clear Mendelian pattern. The disease does run in families, and siblings of diabetic individuals have a 6% risk of developing diabetes compared with 0.4% in the general population. Studies in animals and man have elucidated that the major genetic predisposition is determined by genes within the human major histocompatibility complex (MHC) and in particular is related to DR3, DR4 and DRw9.

The MHC in humans is situated on the short arm of chromosome 6 and covers a region of approximately 3 million bp (Figure 3.4). The products from this region were named human leucocyte antigens (HLA) since they were originally found on human leucocytes. They are divided into three main groups: class I genes at loci A, B and C, class II genes which consist of three different subgroups DP, DQ, DR, and the class III loci which consists of several components of the complement series. Each of these loci has multiple alleles or antigens, each encoding a slightly different gene product. Class I molecules encode peptides which combine with β_2 microglobulin (whose gene lies on chromosome 15) to form class I molecules. Class I molecules are essential to the activation of cytotoxic T lymphocytes, are needed for lysis of cells infected with viruses, and occur on the surface of most nucleated cells. Class II molecules are encoded by three related groups of genes DP, DQ and DR. Each molecule consists of a dimer comprising an alpha and a beta chain. In contrast to the widespread distribution of class I molecules, class II molecules are normally expressed only on specific cell types, notably antigen-presenting cells, B lymphocytes and activated T lymphocytes. Class II molecules activate T helper lymphocytes when presented with foreign antigen on the surface of antigen presenting cells. Class III genes play an important part in clearance of immune complexes and viral neutralisation. Thus, the MHC antigens are critical in mod-

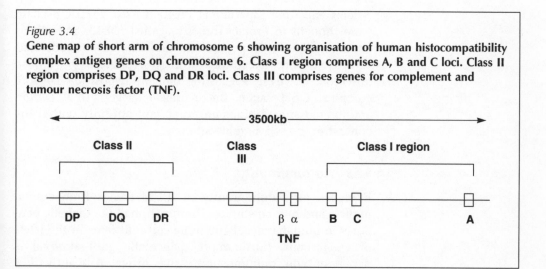

Figure 3.4
Gene map of short arm of chromosome 6 showing organisation of human histocompatibility complex antigen genes on chromosome 6. Class I region comprises A, B and C loci. Class II region comprises DP, DQ and DR loci. Class III comprises genes for complement and tumour necrosis factor (TNF).

ulating the immune response and predisposition to autoimmunity.

Associations with IDDM were first described for HLA-B18 and -B15 and, subsequently, with development of HLA-DR typing for HLA-DR3 and -DR4. There is a higher relative risk associated with DR3 compared to B8 and with DR4 compared with A1, indicating that class II genes are closer to MHC IDDM susceptibility genes than the class I genes. The risk of diabetes to siblings of a diabetic patient is 12% if they are HLA identical.

Considering that MHC is a major genetic factor responsible for controlling the immune response, it is an obvious candidate for the diabetic genetic susceptibility factor. However, the HLA-DR associations are not constant throughout the world. In southern India, although there is an increased prevalance of either DR3 or DR4 in IDDM individuals, there is no additional susceptibility in those individuals with both DR3 and DR4. In Japan, the association is with DR4 and DRw9 and in China with DR3 and DRw9. Furthermore, haplotypes DR3 and DR4 are also common in the general population, 60% of which possess either DR3 or DR4. These, therefore, cannot provide specific prediction markers of IDDM susceptibility.

With advances in molecular biology, the study of associations between HLA and IDDM have been conducted at the DNA level rather than as serological specificities. Studies with subregion specific probes have indicated that associations to the DQ locus were stronger than those observed for DR. DQ alleles are therefore possible determinants of predisposition to IDDM. It has been suggested that particular regions of the DQ genes are crucial to disease susceptibility. Alleles encoding aspartate at position 57 of the DQB chain were thought to protect directly against IDDM, and alleles encoding arginine at position 52 were thought to predispose directly. Several exceptions to these hypotheses have been observed, however, suggesting that predisposition is more complex. On balance, the evidence suggests that certain amino acid substitutions are important but only within the context of the whole molecule.

3.3.3 *Autoimmunity*

Evidence for autoimmunity in IDDM comes from animal models and human studies. Examination of the diabetic pancreas in humans reveals immune cell infiltrates in the islets of Langerhans. Furthermore, pancreatic graft survival is dependent on immunosuppressive drugs. Autoantibodies against specific islet cell antigens have been detected, including autoantibodies directed against glutamate decarboxylase (GAD) and autoantibodies directed against a 38 000 molecular weight (Mw) antigen thought to be an integral part of the insulin's secretory granule membrane. In the non-obese diabetic (NOD) mouse, 80% of females and 20% of males develop diabetes and insulitis develops in all inbred NOD mice. In the BB rat, 60% have diabetes and, furthermore, in the majority of the remaining 40% who do not develop diabetes, there is evidence of islet cell infiltrates and beta cell destruction.

3.3.3.1 Humoral antibodies

There may be a prodromal period of several years before the onset of disease symptoms, and during this period islet cell antibodies, insulin autoantibodies and autoantibodies to specific beta cell proteins may appear in the circulation. These represent markers of beta cell destruction rather than a primary cause. Circulating islet cell antibodies (ICA) are IgG class and are directed against cytoplasmic antigens in beta cells. Spontaneous autoantibodies to insulin are found together with islet cell antibodies during this prolonged prediabetic period, suggesting continuing autoimmune beta cell damage. Other antibodies include autoantigens to a 64 000 Mw islet protein, a 37 000 Mw antigen, probably tyrosine phosphatase 1A-2, a 38 000 Mw antigen and also antibodies to carboxpeptidase H, the enzyme involved in the cleavage of insulin to pro-insulin in the insulin secretory granule. These antibodies may appear in the sera years before the onset of the disease. They are identified most readily in genetically predisposed relatives of people with IDDM. However, not all people who possess any of these antibodies develop diabetes, although those with two or three antibodies appear to be at a higher risk of developing diabetes compared with the general population.

The clinical relevance of these antibodies is indicated in the assessment of risk of IDDM developing in first-degree relatives. In a recent family study, relatives with ICA only had a risk of progression to diabetes of 80% in 10 years. In those who also had insulin antibodies, the risk was 84% compared to 61% when combined with GAD antibodies. Islet cell antibodies occurring together with a 37 000 Mw antibody gave a risk of 76% and was associated with rapid progression of the disease. The presence of three or more antibodies increased the risk to 88%. ICA may be directly cytotoxic to islets in culture, but since islet cell function is normal in some patients with ICA they might only represent markers of damage from some other cause.

3.3.3.2 Cellular immunity

Apart from possible circulating antibodies, different classes of inflammatory cells may also be important. T lymphocytes are primarily involved in cell-mediated immune responses and can be subdivided into several different types.

T helper lymphocytes are activated by being exposed to a foreign antigen together with an HLA class II antigen on the surface of an antigen-presenting cell. Activated T helper lymphocytes release peptides which can activate macrophages and cytotoxic T lymphocytes.

Cytotoxic T lymphocytes are activated by the combination of foreign antigen, particularly viruses, and HLA class I molecules present on the surface of cells and then lyse these target cells. Cytotoxic T lymphocytes carry a specific CD8 antigen, and T helper cells the CD4 antigen.

T suppressor lymphocytes inhibit cell-mediated and humoral immune responses. They are apparently activated by T helper lymphocytes and act to inhibit the latter, providing a 'damping' circuit which prevents an excessive immune response.

Insulitis (the mononuclear cell infiltration of islets) consists mainly of cytotoxic/suppressor T lymphocytes and activated T lymphocytes. The presence of these cells together with penetration of IgG into islet beta cells and local complement deposition indicate an autoimmune process, leading to destruction of beta cells only.

In humans, cellular immunity has been investigated by studying peripheral lymphocytes which have shown a marked increase in the number of activated lymphocytes in the blood of patients with newly diagnosed IDDM. An alternative approach is the use of animal models including the BB (or biobreeding) rat. This mutant animal develops diabetes after 60–120 days, associated with marked insulitis, hyperglycaemia and ketosis. Transfer of lymphocytes from afflicted to normal rats results in transfer of diabetes.

Cytokines released by activated lymphocytes and macrophages may enhance autoimmune damage either by direct islet beta cell toxicity or by inducing aberrant or enhancing normal expression of HLA molecules on islet beta cells. In twin studies, TNF (tumour necrosis factor) and IL-1 (interleukin 1) levels were elevated above the normal range in 70% of those twins who developed IDDM subsequently.

However, the twins who remained non-diabetic had elevated levels in less than 15% of samples.

3.3.4 *Environmental factors*

These fall into two groups: viruses and toxic chemicals. Alloxan and streptozotocin are highly diabetogenic chemicals but there is no evidence to incriminate these as a cause of diabetes in man, although there has been an indication from Iceland that smoked mutton, which contains toxic *N*-nitroso compounds, may cause diabetes. Streptozotocin is a nitroso derivative.

Viruses may damage beta cells by direct invasion or by triggering an autoimmune response. They also persist within the beta cells causing long-term interference with their metabolic and secretory function. Coxsackie beta viruses, especially B4, can cause acute pancreatitis and beta cell destruction, and some serological studies suggest exposure to the virus in newly diagnosed IDDM patients. Other viruses implicated in human IDDM include echoviruses, cytomegalovirus and herpesvirus. There are a large number of experimental models in which diabetes can be produced by viruses in animals. Twenty viruses are known to attack the islets of Langerhans in animals and prominent among these are the small RNA viruses of the picorna group.

Aberrant HLA class II antigen expression on islet beta cells may be induced by viruses and other environmental agents. Beta cells can then act as antigen-presenting cells, exposing their own surface antigens to T helper lymphocytes.

In view of these immunological changes, there have been studies to prevent the onset of diabetes by immunosuppression and these have been successful in the diabetes of the BB rat. In humans, initial attempts with steroids, with or without azathioprine, were not successful, but cyclosporin A has produced remission in a small proportion of patients.

In conclusion, the immune response during the prediabetic period involves both cellular and humoral changes, probably initiated by an environmental factor and maintained by islet cell antigens. A combination of environmental and genetic factors which trigger an autoimmune attack on the beta cells seems likely to be responsible for beta cell destruction during the onset of diabetes in genetically susceptible individuals.

3.4 Molecular genetics and aetiology of non-insulin-dependent diabetes (NIDDM)

3.4.1 Molecular genetics

NIDDM is much more heterogeneous in its pathogenesis and clinical presentation than IDDM. NIDDM clusters in families and its high concordance rates in monozygotic twins indicate the presence of important predisposing genetic factors. Many of the apparently unaffected co-twins of the diabetic twins have subclinical defects in insulin secretion. In contrast to IDDM, there are no strong associations between NIDDM and HLA types in caucasian populations.

Different genetic defects are associated with an NIDDM-like phenotype, as indicated by the observation that NIDDM associates with a variety of genetically different congenital diseases such as ataxia telangiectasia, Praeder–Willi syndrome and Werner's syndrome. These observations suggest that NIDDM may be composed of different subtypes, each with distinct genetic defects and involvement of different pathophysiological mechanisms. However, despite intensive research and study, the genetic defects and corresponding pathobiological mechanisms which contribute to the development of NIDDM are not understood in the majority of patients.

However, particular diabetic subtypes have recently been associated with distinct genetic lesions; these are shown in Box 3.1.

3.4.2 Genetic defects of the beta cell

3.4.2.1 Maturity-onset diabetes of the young (MODY)

MODY is characterised by the onset of mild hyperglycaemia at an early age and is inherited in an autosomal dominant pattern. Patients with MODY have an impaired insulin secretion rather than a decreased insulin action. Abnormalities have been identified at three genetic loci on different chromosomes. Mutation in the glucokinase gene on chromosome 7p results in a defective glucokinase molecule. Glucokinase converts glucose to glucose 6-phosphate, the metabolism of which in turn stimulates insulin secretion by the beta cell. Thus, glucokinase acts as the glucose sensor for the beta cell. As a result of these defects in the glucokinase gene, increased levels of glucose are necessary to produce normal levels of insulin secretion. The second lesion has been identified on chromosome 20q and the third lesion on chromosome on 12q. The mechanisms by which the latter two abnormalites

Box 3.1 **Association of diabetic subtypes with specific genetic lesions**

A. Patients with known genetic defects of the beta cell
   1. Maturity-onset diabetes of youth (MODY)
      a. Chromosome 7 (glucokinase defects)
      b. Chromosome 20
      c. Chromosome 12
   2. Insulinopathies
      a. Pro-insulin abnormalities
      b. Insulin chain mutations
   3. Mitochondrial defects often associated with sensory deafness
B. Patients with genetic defects in insulin action
   1. Alterations in insulin receptor structure and function
      a. Type A insulin resistance
      b. Leprechaunism
      c. Rabson–Mendenhall syndrome
      d. Others
   2. Alterations in postreceptor signal transduction
      a. Polycystic ovary syndrome
      b. Lipoatrophic diabetes
      c. Patients with anti-insulin receptor antibodies

cause impaired insulin secretion and hyperglycaemia are unknown.

3.4.2.2 *Insulinopathies*

Insulinopathies arise from rare mutations in the human insulin gene which lead to the synthesis and secretion of abnormal gene products. Genetic abnormalities that result in the inability to convert pro-insulin to insulin have been identified and these traits are inherited in an autosomal dominant pattern. The resulting carbohydrate intolerance is mild. Furthermore, the production of mutant insulin molecules with resultant impaired receptor binding has been identified and is associated with autosomal inheritance and mildly impaired carbohydrate metabolism. These patients display degrees of glucose intolerance but a normal response to exogenous insulin administration. The defect is inherited in an autosomal fashion.

Insulin gene mutations fall into two groups: those that interfere with the processing of pro-insulin to insulin and those leading to the secretion of abnormal insulins with molecular weights close to that of normal insulin. Mutations leading to pro-insulins Tokyo and Boston give rise to the secretion of intermediates of pro-insulin processing in which the C-peptide remains joined to the insulin A chain. These products contain the normal structure of insulin but this remains cryptic since complete conversion of the precursor to insulin is not possible. Insulins Chicago, Los Angeles and Wakayama result from mutations from the second group. In each case, a single nucleotide change in the insulin gene has resulted in the synthesis of an abnormal insulin bearing an amino acid replacement at a site known to be important for interaction with the insulin receptor.

All the abnormal insulins exhibit severely decreased receptor binding and biological potencies. They are not rapidly cleared from the blood because the mechanisms for insulin degradation are crucially dependent on the initial binding of the hormone to receptors and on subsequent receptor-mediated events. Thus, affected subjects are hyperinsulinaemic as a result of the slow degradation of abnormal forms. However, insulin gene mutations are rare within the human population. The diabetes is mild and is never insulin-dependent.

All mutations discovered so far are point mutations. A study of these mutant human insulins has provided very important information about insulin structure–function relationships. The substitutions so far discovered fall into three classes, with the following implications on structure–function relationship:

1. The first identifies the importance of the basic amino acid pair Lys-Arg, which links the C-peptide to the insulin A chain in the conversion of pro-insulin.

2. The second identifies the importance of three residues of the insulin molecule, A3 Val, B24 Phe and B25 Phe, in directing the affinity of the insulin–receptor interactions.

3. The third class identifies a special case in suggesting the importance of B10 His in directing both pro-insulin processing and insulin interactions with its receptor.

3.4.2.3 *Mitochondrial defects associated with sensory deafness*

A point mutation in mitochondrial DNA is maternally transmitted and is associated with diabetes and deafness. This occurs at position 3243 in the tRNA leucine gene, leading to an adenine-to-guanine transition mutation.

3.4.2.4 *Wolfram's syndrome*

Wolfram's syndrome is an autosomal recessive disorder characterised by insulin-deficient diabetes and the absence of beta cells. It is characterised by diabetes insipidus, diabetes mellitus, optic atrophy and deafness.

3.4.3 *Genetic defects in insulin action*

In the majority of patients with NIDDM, the insulin receptor gene is normal both in populations with a high occurrence of this disease as well as in Caucasians with the common form of NIDDM. There is a subset of diseases in which pathogenesis is closely related to abnormalities of the insulin receptor function. The syndromes have been designated type A, type B and their variants. The type A syndrome appears to be due to genetic abnormalities in the insulin receptor, whereas the type B syndrome is associated with anti-insulin receptor autoantibodies. Thus, type B is strictly not an alteration in insulin receptor structure and is now classified separately. Variants of the type A syndrome with normal insulin receptors and postreceptor defects have occasionally been called type C syndrome.

3.4.3.1 *Alterations in insulin receptor structure and function*

The type A syndrome, resulting from various genetic defects in the insulin receptor, predominantly affects young women, who are grossly hyperinsulinaemic, markedly glucose intolerant and usually virilised. Some individuals with these mutations have acanthosis nigricans. Various mutations have been described. In one study there was a single nucleotide substitution: 735 AGG to AGT causing a single amino acid substitution which blocked the proteolytic cleavage sites. In another study, eight mutations in the receptor gene sequence were noted. Six did not alter the amino acid sequence but two were nonsense mutations which would prematurely terminate transcription of the alpha-subunit and delete both the transmembrane and tyrosine kinase domains of the beta-subunit. In type A syndrome, many patients exhibit reduced insulin binding resulting either from reduced receptor number or from altered receptor infinity. Others have been shown to have defective receptor autophosphorylation or abnormal post-translational processing of the receptor precursor.

Leprechaunism and the Rabson–Mendenhall syndrome are two paediatric syndromes that have mutations in the insulin receptor gene with subsequent alterations in insulin receptor function and extreme insulin resistance. These and other patients with alterations in insulin receptor function may have a defect in any part of the sequence of receptor synthesis, transport of the receptor to the plasma membrane, binding of the receptor to the insulin molecule, transmembrane signalling, and recycling or degradation of the receptor after endocytosis.

3.4.3.2 *Alterations in postreceptor signal transduction*

Alterations in the structure and function of the insulin receptor cannot be shown in patients with polycystic ovary syndrome or lipoatrophic diabetes. Thus, it is probable that the lesions are present in the postreceptor signal transduction pathway.

3.4.4 *Anti-insulin receptor antibodies*

In the past, this syndrome was termed type B syndrome. It is due to antibodies, usually IgG, directed against the insulin receptor. It mainly affects women, who often have other features of generalised autoimmune disease. These antibodies cause diabetes by binding to the insulin receptor and thereby reducing the binding of insulin to target tissues. Most patients have hyperglycaemia, but specific receptor stimulating antibodies may cause hypoglycaemia. The precise aetiology of this syndrome is not known but is apparently due to a primary defect in the immune system rather than in the insulin receptor itself. Glucose intolerance is a minor component of the disease, but careful follow-up is necessary as several patients have progressed from severe insulin resistance to profound hypoglycaemia.

3.4.5 *Aetiology of NIDDM*

The beta cell dysfunction in patients with NIDDM is manifested as the loss or blunting of the acute first phase and early insulin release to glucose stimulation. The mechanism of the glucose unresponsiveness by the beta cells in patients with idiopathic NIDDM is unknown. Several studies have linked the abnormality to the transport and subsequent metabolism of intracellular glucose by the beta cells. The enzyme glucokinase is the rate-limiting step that catalyses ATP-dependent phosphorylation of glucose to glucose 6-phosphate. Recent studies in patients with MODY or early-onset NIDDM have identified genetic mutations in the glucokinase gene in some patients (see above). Studies indicate that alterations in the glucokinase gene may be important in the pathogenesis of early-onset diabetes or NIDDM. This may be racially and ethnically dependent. However, its role in idiopathic adult-onset NIDDM remains debatable.

Total islet and beta cell mass is reduced to 50–60% of normal in NIDDM. The most common morphological feature is amyloid deposition in the islets. These deposits consist of IAPP (islet amyloid polypeptide) or amylin. Amylin is co-segregated in the beta cells as well as co-secreted with insulin C-peptides. Genetic mutation appears to be an unlikely cause of the increased amylin in NIDDM patients. The theory that overproduction of IAPP is a major cause of insulin resistance in NIDDM is unproven, and it has not been incontrovertibly shown that IAPP is of importance for the development of any of these abnormalities in humans.

The role of insulin resistance and the aetiology of the hyperglycaemia in NIDDM is now established. The insulin resistance is found predominantly in the insulin-sensitive tissues, namely the liver, adipose tissue and skeletal muscle. Most patients with idiopathic NIDDM have predominantly postreceptor insulin abnormalities that are responsible for the alterations in glucose disposal, although few patients with insulin receptor defects due to mutant genes have been reported. However, the actual genetic defect responsible for insulin resistance remains unknown.

3.5 The molecular basis of diabetic complications

Complications of diabetes can be divided into two groups: microvascular, affecting eyes, kidneys and nerves, and macrovascular, affecting the cororary, cerebral and peripheral vascular systems. The various clinical syndromes share a common pathophysiological feature, namely a progressive narrowing of vascular lumina leading to inadequate perfusion of target organs.

The microvascular complications of diabetes include retinopathy, nephropathy and neuropathy, although the contribution of microangiopathy to neuropathy remains uncertain. Prolonged exposure to elevated glucose concentrations damages tissues, but the mechanisms are not fully understood. Two main processes have received intensive study. These include the polyol pathway and glycation of proteins, which initially leads to the formation of reversible compounds. This is followed by cumulative irreversibe changes in long-lived molecules by the formation of advanced glycation end-products on matrix proteins such as collagen and on nucleic acids and nucleoproteins.

The basic mechanisms of these complications will be considered first and then the pathogenesis in the affected organ systems described.

3.5.1 *Polyol pathway*

In insulin-independent tissues such as nerve, glomerulus, lens and retina, hyperglycaemia causes elevated tissue glucose levels. The enzyme aldose reductase catalyses the reduction of glucose to sorbitol, which is subsequently converted to fructose (Figure 3.5). Sorbitol accumulates in the cell, causing damage by osmotic effects and altered redox state of pyridine nucleotides. NAD^+ is reduced to NADH by sorbitol dehydrogenase, leading to an increased ratio of $NADH/NAD^+$.

3.5.2 *Glycation*

When glucose attaches to an amino group this leads to glycation or Schiff base products. Two factors determine the extent of early glycosylation product formation in vivo, namely glucose concentration and the duration of the exposure to glucose. The Schiff base adducts then undergo Amadori rearrangement to form stable ketoamines known

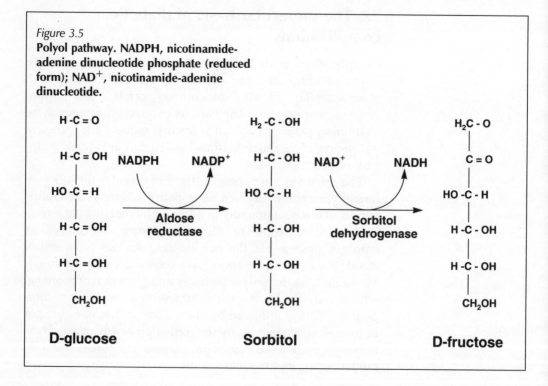

Figure 3.5
Polyol pathway. NADPH, nicotinamide-adenine dinucleotide phosphate (reduced form); NAD^+, nicotinamide-adenine dinucleotide.

as Amadori products. These products then undergo further rearrangements and fragmentation to form highly reactive carbonyl compounds which react with amino groups to form stable, irreversible glycated end-products. Such glycation may affect the function of a number of proteins and be partly responsible for free-radical-mediated damage in diabetes. In long-lived molecules, early glycation products slowly and irreversibly form cross-linkings to form advanced glycation end-products (AGE). Glucose-derived AGEs result from covalent cross-linking of protein molecules which follows one of two patterns. One type closely resembles the imidazole-based glucose-derived cross-links and appears to form from the condensation of two Amadori products (Figure 3.6). The other pattern of AGE cross-linking results from the reaction of an Amadori product with the Amidori-derived compound 3-deoxyglucosone. This highly reactive dicarbonyl compound cyclizes to form pyrrole intermediates which then react with amino groups to form pyrrole-based cross-links (Figure 3.7).

In addition, AGEs elicit a wide range of cell-mediated responses, leading to vascular dysfunction, matrix expansion and athero- and glomerulosclerosis. Cellular responses are thought to be largely induced through an AGE-specific cell-surface receptor complex that has been identified on many cell types.

Figure 3.6
Formation of imidazole-based glucose-derived cross-links. Glucose attaches to an amino group of a protein to form a Schiff base. The Schiff base undergoes Amadori rearrangement to form stable products. Condensation of two Amadori products leads to imidazole-based glucose-derived cross-links.

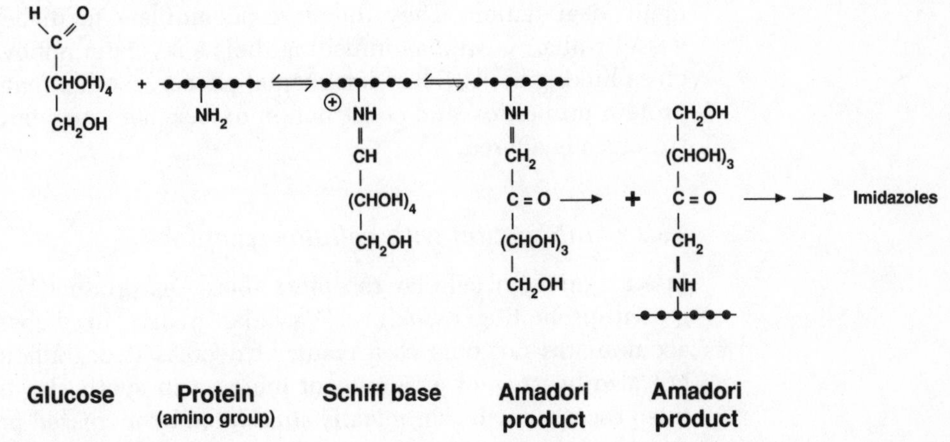

Figure 3.7
Formation of pyrrole-based cross-links from the reaction of an Amadori product with the Amadori-derived compound 3-deoxyglucosone.

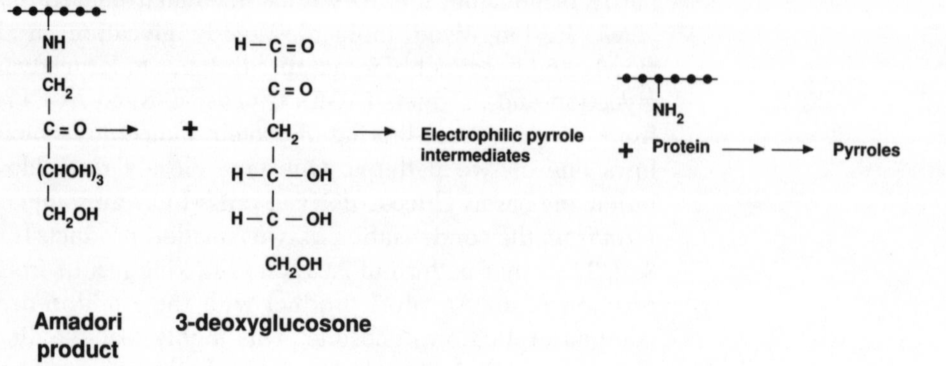

Amadori product   3-deoxyglucosone

3.5.3 *Pathological consequences of advanced glycosylation product (AGE) formation*

A number of the irreversible AGEs are capable of forming covalent bonds with nearby amino groups on other proteins and nucleotides, resulting in glucose-derived cross-links. This could contribute to diabetic tissue damage, including effects on extracellular matrix proteins, specific cellular receptors or nucleic acids and nucleoproteins.

3.5.3.1 *Extracellular protein cross-linking*

Extracellular protein cross-linking irreversibly traps deposited plasma proteins and reduces their susceptibility to enzymatic degradation. They therefore accumulate in diabetic vessel walls. As well as impeding their enzymatic removal, cross-linking by AGE has detrimental effects on other matrix protein properties, and construction of basement membrane structure is altered.

3.5.3.2 *Interaction with cellular receptors*

Interaction with cellular receptors increases production of growth-promoting cytokines. Vascular matrix in diabetes accumulates not only as a result of decreased degradation but also because of a significant increase in synthesis, and this process may be chronically stimulated by increased production of growth promoting factors. Macrophages are known to carry a high-infinity receptor for AGE proteins. Interaction with cellular receptors increases production of growth-promoting cytokines that augment matrix synthesis, stimulate hypertrophy and hyperplasia and induce procoagulatory changes in the endothelial surface.

3.5.3.3 Intracellular nucleic acid cross-linking

Formation of AGE on DNA is associated with mutations and altered gene expression. Human endothelial cells cultured in 30 mmol l^{-1} glucose display an increase in single-stranded DNA breaks and in DNA repairs. There is also a decrease in transcriptional regulatory protein binding.

3.5.3.4 Relationship of AGE formation to other pathogenetic factors

Increased intravascular pressure accelerates the development of AGE-induced pathological changes by increasing extravasation of plasma proteins. Excessive AGE formation promotes thrombogenic changes at the endothelial surface by stimulating TNF and IL-1 secretion by macrophages. These changes generate thrombin and activator factor 10A which stimulate the release of platelet-derived growth factor which in turn accelerates both hyperplasia and hypertrophy in the diabetic vessel wall. Macrovascular disease is also an important problem in diabetes, although plasma LDL levels are not always abnormal. However, accumulation of AGE on arterial wall collagen would enhance extracellular deposition of lipoprotein. This would act as a nucleus for further AGE formation, and interaction with cellular AGE receptors would stimulate growth factor release.

As AGE formation is implicated in the pathogenesis of chronic diabetic complications, pharmacological agents have been investigated to inhibit this process. Aminoguanidine selectively blocks reactive carbonyl groups on early glycosylation products and on their derivatives such as 3-deoxyglucosone. This effectively inhibits the formation of AGE and blocks the AGE cross-linking of soluble proteins to matrix and of collagen. In animal studies, it reduces the matrix content of AGE and of cross-linked plasma proteins in both aorta and kidney.

3.5.4 Retinopathy

Microaneurysms and vascular leakage are characteristic of early diabetic retinopathy while frank neovascularisation (retinal angiogenesis) occurs later in response to increasing capillary occlusion and retinal ischaemia. Angiogenic growth factors released locally from ischaemic retina are essential for the induction of new vessels, and several specific growth factors have been identified in association with retinal cells including insulin-like growth factor-1, fibroblast growth factors, transforming growth factor-β and, more recently, vascular endothelial cell growth factor.

Increased polyol pathway activity in pericytes generates sorbitol under hyperglycaemic conditions. Accumulation of sorbitol is associated with basement membrane thickening and pericyte loss. Endothelial cell proliferation and then degeneration occur and capillary closure is associated with disappearance of the endothelial cell. There may be an early increase of retinal flow and accumulation of sorbitol in the vessel wall. Capillary closure and retinal ischaemia result in excessive leakage from diseased capillaries and ischaemic areas provide the stimulus for the growth of new vessels.

3.5.5 *Nephropathy*

The hallmark of nephropathy is albuminuria, although only 30–40% of IDDM patients develop this complication. Albumin is a negatively charged molecule and albuminuria is caused by a loss of glomerular charge selectivity which is essentially determined by the presence of negatively charged heparan sulphate residues. The pathogenetic event that is crucial for the development of an albuminuria is a decreased concentration of heparan sulphate. Hyperglycaemia depresses the activity of glucosamyl-*N*-deacetylase, an important enzyme in heparan sulphate biosynthesis. Genetically determined variations in the susceptibility of this enzyme to inhibition by hyperglycaemia may explain the subset of IDDM patients and nephropathy. Albuminuria is associated with premature atherosclerosis. Several studies have demonstrated a highly significant negative correlation between heparan sulphate and atherosclerosis in human arteries. Furthermore, in diabetic patients, there is a 50% reduction in acid glycosaminoglycans within the media of extramural coronary arteries.

3.5.6 *Neuropathy*

Both metabolic and vascular factors are important in the pathogenesis of neuropathy. Hyperglycaemia can lead to glycation of neural proteins with a subsequent decrease in axonal transport and diminished nerve conduction. Elevated blood glucose can competitively inhibit the nerve uptake of myoinositol, a neural membrane component. Decreased neuromyoinositol has been associated with decreased nerve function.

Morphometric assessment of perineural capillaries have recently shown abnormalities due to endothelial cell hypertrophy and hyperplasia. These result in reduced transperineural and endoneural blood flow, which leads to endoneural hypoxia. In the diabetic nerve, axonal dwindling and segmental demyelination associated with changes in the vasa vasorum produce motor, sensory and autonomic dysfunction.

3.5.7 *Macrovascular disease*

Large vessel disease is a leading cause of mortality in diabetes and a major cause of morbidity. Numerous autopsy studies of coronary arteries have also shown that, relative to non-diabetics, the arteries of diabetics shown more evidence of atherosclerosis, including more fatty streaks, more raised atherogenic lesions, more fibrous plaques and a greater extent of coronary stenosis. The major abnormalities of the macrocirculation are characterised by premature atherosclerosis with accumulation of lipoproteins within the vessel wall.

Non-enzymatic glycation could affect early atherogenesis by several mechanisms. Early glycation products stimulate the generation of free radicals and AGE proteins which are chemotactic for monocytes. Binding of AGE proteins to specific receptors induces the production of several cytokines. Binding of AGE protein to endothelial cells results in procoagulatory changes on the surface of these cells. AGE can also impair the binding of heparan sulphate to the extracellular matrix, resulting in a loss of negative anionic sites thereby increasing endothelial permeability. Furthermore, AGE can quench the endothelium-derived nitric oxide which could lead to abnormal endothelium-dependent vasodilatation.

Abnormalities of lipoproteins are strong predictors of cardiovascular disease. Oxidation of LDL may be important, and oxidative modification of LDL or other lipoproteins occurs in vivo in the artery wall in experimental animals and man. In diabetes there is enhanced susceptibility to oxidation of LDL, and modification of these lipoproteins within the intima of the vessel wall is important. In the circulation, LDL is protected by circulating antioxidants but, once trapped in the intima, the polyunsaturated fatty acids in LDL can be

attacked by free radicals resulting in the formation of oxydised LDL. This can result in stimulation of macrophages within the vessel wall. Furthermore, oxidised LDL is cytotoxic for endothelial cells. Hyperglycaemia also affects endothelial function resulting in increased permeability, altered release of vasoactive substances, increased synthesis of extracellular matrix components, increased production of procoagulation proteins and decreased production of fibrinolytic factors.

4 Haemoglobinopathies: A Paradigm of Molecular Disease

D. Mark Layton

4.1 Introduction

The discovery by the American chemist Linus Pauling and his colleagues in 1949, that sickle cell anaemia results from an abnormal haemoglobin molecule, signalled the dawn of the molecular era in medicine. Nearly a decade later, Ingram identified the chemical change in sickle haemoglobin which accounted for the altered electrical charge Pauling had demonstrated by electrophoresis as a single amino acid substitution of valine for glutamic acid at position 6 of the β-globin polypeptide chain. At about the same time, after many years effort, Perutz resolved the structure of the haemoglobin molecule which unlocked the mystery of how haemoglobin fulfils its role in respiratory transport by switching between two alternative structures. These scientific landmarks held profound implications for molecular biology, genetics and protein chemistry. Between 1954 and 1962, in recognition of this, Pauling, Ingram and Perutz were each awarded the Nobel prize for their work on the function and pathology of haemoglobin.

With the advent of recombinant DNA technology in the early 1970s, the genes for α- and β-globin were among the first human genes to be cloned. This set the stage for identification of the molecular defects responsible for human haemoglobin disorders. The picture to emerge has provided an almost complete portrayal of the repertoire of mutations that underlie single gene disorders and shaped our understanding of the complex interaction between genotype and phenotype in human genetic disease.

4.2 Global distribution and the malaria hypothesis

Worldwide, the major haemoglobin disorders sickle cell and thalassaemia are, together with glucose 6-phosphate-dehydrogenase deficiency, the most common inborn errors in man. An estimated 250 million people carry a haemoglobinopathy gene and at least 200 000 affected homozygotes are born annually. It is predicted that, by the year 2000, 7% of the world's population will carry a gene for sickle cell or thalassaemia. In parts of equatorial Africa the prevalence of sickle cell trait reaches 30%, and similar frequencies for thalassaemia can be found in the Mediterranean, Asia and Africa.

The geographical distribution of haemoglobinopathies bears a striking resemblance to that of malaria, one of the most powerful selective factors affecting human populations. As first proposed by Haldane in 1948, and now supported by a substantial body of experimental and epidemiological evidence, this reflects the protection afforded to carriers against *Plasmodium falciparum*. Haemoglobinopathies thus represent an example of a balanced polymorphism in which increased reproductive fitness among heterozygotes outweighs the survival disadvantage of homozygotes.

4.3 The genetic control of haemoglobin synthesis

Each haemoglobin molecule is made up of four polypeptide chains, two α-like chains of 141 amino acids and two β-like chains of 146 amino acids. The synthesis of α- and β-like chains is normally balanced carefully to allow correct tetramer assembly. To each peptide chain is attached a molecule of the ferroporphyrin haem which gives blood its red colour. At the centre of each haem group lies an atom of iron through which oxygen is bound. Allosteric modulation of this structure serves to allow reversible binding of oxygen.

The haemoglobin (Hb) composition of red cells varies during development. In humans, two switches take place, the earliest at 6–8 weeks of fetal life from embryonic (Hb Gower-1 $\zeta_2\varepsilon_2$, Hb Gower-2 $\alpha_2\varepsilon_2$, Hb Portland $\zeta_2\gamma_2$) to fetal Hb F $\alpha_2\gamma_2$ forms with a second transition from fetal to adult (Hb A $\alpha_2\beta_2$) synthesis perinatally. A second minor haemoglobin called Hb A$_2$ $\alpha_2\beta_2$ persists throughout adult life. In normal individuals by 1 year of age at least 95% of the haemoglobin in red cells is Hb A, less than 3.5% Hb A$_2$, and less than 1% Hb F. How the development switch in haemoglobin synthesis is determined is still unclear. It appears to be largely independent of environmental factors suggesting there is some form of 'biological clock' preprogrammed into haemopoietic cells.

The change in the pattern of haemoglobin synthesis during development has important implications for the clinical expression of haemoglobin disorders. Since α chains are shared by both fetal and adult haemoglobin, in the most severe form of α-thalassaemia where the capacity for their synthesis is lost completely no functional haemoglobin can be formed beyond the embryonic stage. The severe anaemia which results leads to fetal hydrops and almost invariably intrauterine death or stillbirth. By contrast, the clinical manifestation of β chain disorders is delayed until the fetal-to-adult switch is complete after birth.

Human haemoglobin genes are encoded by two loci, the α-like genes (ζ, $\alpha2$ and $\alpha1$) on chromosome 16 and the β-like genes (ε, γ, δ, and β) on chromosome 11. The array of genes reflects the order in which they are expressed during development. Each globin gene comprises three exons (coding regions) and two introns (non-coding regions) and spans 1–2 kb. The different globin genes show considerable homology and probably evolved from a common ancestral gene. At the β locus there are two fetal genes, $G\gamma$ and $A\gamma$, which code for almost identical proteins differing by only a single amino acid (these arose by gene duplication). During development, the relative expression of the two fetal γ genes varies, $G\gamma$ predominating in fetal life and $A\gamma$ postnatally. Transcription of globin genes requires interaction between their promoter and other key regulatory elements and both ubiquitous and erythroid specific DNA binding proteins (*trans*-acting factors).

4.4 The thalassaemias: quantitative defects in haemoglobin synthesis

The thalassaemias are the most common human autosomal recessive disorders. The most important and clinically significant forms are α- and β-thalassaemia and the thalassaemia-like structural variant Hb E. These may be subdivided according to whether the underlying molecular defect abolishes completely (α^0- or β^0-thalassaemia) or reduces (α^+- or β^+-thalassaemia) globin production. In the case of α-thalassaemia this is complicated by duplication of the α genes (see above), which provides a greater reserve against loss of output and generates a spectrum of phenotypes ranging from clinically silent to lethal as the capacity for a chain synthesis is progressively lost.

The pathophysiology of thalassaemia can be understood in terms of the degree and consequences of imbalance in globin synthesis. In homozygous α^0 thalassaemia little or no functional haemoglobin can be made after early fetal life. Surplus fetal γ chains combine to form tetramers (γ_4) recognised electrophoretically as Hb Bart's which is ineffective as an oxygen transporter. The resulting anaemia and tissue hypoxia lead to fetal hydrops. Even a small amount of residual α chain synthesis is sufficient to allow some fetal and adult haemoglobin to be produced. Conversely, in homozygous β-thalassaemia, the pathogenesis of anaemia reflects ineffective erythropoiesis due to the accumulation of surplus α chains, which precipitate in red cell precursors leading to their destruction in the bone marrow. In a futile attempt to compensate for this there is massive expansion of erythropoiesis in the bone marrow and extramedullary sites.

Over the past two decades, remarkable progress has been made towards unravelling the molecular basis of thalassaemia. This has unveiled a picture of extraordinary heterogeneity. In 1974 Hb Bart's hydrops fetalis was shown to result from a mutation involving both α-globin genes, the first example of a human gene deletion. Subsequently, several other mutations which delete both α genes have been discovered. These include very large deletions that remove the entire α-globin complex (including both ζ and α genes) which have only been observed in the heterozygous state, presumably because in homozygotes they are lethal in early development. The 5′ prime and 3′ deletion breakpoints of several α^0-thalassaemia mutations cluster, suggesting there may be regions of the α-globin complex prone to recombination. In some cases this appears to be mediated by recombination between *Alu* repeat sequences. Several other mechanisms, including terminal truncation of chromosome 16 and deletion of the major regulatory element HS-40, have also been shown to cause α^0-thalassaemia. Although rare, these examples are of considerable interest because of their potential relevance to the molecular pathology of other genetic diseases whose loci lie at the telomeric end of chromosomes or come under the influence of a locus control region.

Whilst α^0-thalassaemia is largely restricted to Chinese, South-East Asian and some Mediterranean populations, α^+-thalassaemia is more widely distributed throughout the Mediterranean, Africa and Asia. The molecular pathology of α^+-thalassaemia teaches us other important lessons. Each α gene is embedded within a region of homology 4 kb long comprising homologous subsegments X, Y and Z. Misalignment and crossover during meiosis generates chromosomes which carry either single ($-\alpha$) or triplicated ($\alpha\alpha\alpha$)

genes. Homologous recombination between Z boxes deletes 3.7 kb and X boxes 4.2 kb of DNA, accounting for the two most common α^+-thalassaemia mutations $-\alpha3.7$ and $-\alpha4.2$. Several non-deletional mutations which cause α^+-thalassaemia have also been identified. Most of these involve the dominant $\alpha2$. This bias probably reflects the more pronounced phenotype and perhaps selective advantage. Several non-deletional forms of α^+-thalassaemia (e.g. Hb Constance Spring) result from point mutations which replace the translation termination codon TAA by an amino acid allowing normally untranslated mRNA to be read in-frame until the next stop codon is reached. This produces an elongated α chain which results in a structurally abnormal haemoglobin and also reduces the stability of mRNA accounting for the thalassaemic phenotype.

In contrast to α-thalassaemia, the majority of molecular defects which underlie β-thalassaemia are point mutations. The level at which these disrupt gene expression varies. To date, over 120 different mutations have been identified, although in most populations, as a consequence of their independent origin, only a few mutations are common. This allows a rational approach to molecular diagnosis of β-thalassaemia in different ethnic groups.

4.5 Prenatal diagnosis

The ability to diagnose haemoglobinopathy in fetal life by testing for the specific gene defect (S) carried by both parents or RFLP (restriction fragment length polymorphism) linkage (see p. 15) opened the way for prenatal diagnosis in the first trimester of pregnancy. This offers families at risk of having children with severe forms of thalassaemia or sickle cell disease the option to avoid the birth of an affected child. Analysis of DNA obtained by chorionic villus sampling may be performed from as early as 11 weeks of gestation. The polymerase chain reaction (PCR) enables prenatal diagnosis to be accomplished rapidly (usually within 3 days). Where a linked RFLP is used to track the disease allele, potential error due to meiotic recombination should be incorporated into prediction of fetal genotype (see p. 10). The emergence of new technologies in reproductive medicine may in the future open the possibility of genetic diagnosis before implantation. Applying powerful PCR techniques, a single cell removed from the embryo days after fertilisation in vitro can be tested and only embryos shown to be unaffected selected for transfer. Providing implantation takes place successfully, this ensures the pregnancy will not be affected. Although limited currently by the low pregnancy rate after in vitro fertilisation, preimplantation genetic diagnosis may broaden the reproductive options for families at risk of having children with a major haemoglobinopathy, particularly those for whom conventional prenatal diagnosis and elective abortion remain unacceptable on religious or other ethical grounds.

4.6 Future prospects

The past two decades have witnessed remarkable improvements in the outlook for patients with haemoglobinopathies in the developed world. The introduction of desferrioxamine to prevent iron overload and death due to myocardial siderosis has prolonged survival in homozygous β-thalassaemia. Similarly, in sickle cell disease, newborn screening and the adoption of preventive measures have substantially reduced the incidence of life-threatening pneumococcal sepsis and acute splenic sequestration, formerly major causes of fatality in early childhood. The correction of thalassaemia major and sickle cell disease remains a more elusive goal. Bone marrow transplantation offers the prospect of cure but, though successfully applied in both β-thalassaemia major and more recently sickle cell anaemia, is only applicable to the minority of cases who have an HLA-matched donor within the family. The search for more widely applicable strategies for the treatment of haemoglobinopathies currently focuses on two novel approaches.

Reversal of the developmental switch from fetal to adult haemoglobin synthesis represents an attractive approach to therapy of β chain disorders. Whereas in β-thalassaemia relatively high levels of fetal haemoglobin (Hb F) would be required to restore balanced chain synthesis because the production of fetal γ and sickle β chains is regulated in a reciprocal manner, and Hb F inhibits familiarisation of sickle haemoglobin, significant clinical benefit may be gained from more modest Hb F increments in sickle cell disease. Several pharmacological agents, including cytotoxic drugs such as hydroxyurea and short chain fatty acids, have been shown to reactivate Hb F synthesis. The latter were discovered following the observation that haemoglobin switching was retarded in the infants of diabetic mothers due to high plasma falls of butyrate. Early clinical trials of both hydroxyurea and butyrate compounds have shown promising results.

5 Lipids, Lipoproteins and Atherosclerosis

William J. Marshall

5.1 Introduction

Four major classes of lipids circulate in the blood: cholesterol and cholesteryl esters, triacylglycerols (colloquially, and more frequently, referred to as 'triglycerides'), phospholipids and free (i.e. non-esterified) fatty acids. All are essential to the normal function of the body. All are poorly soluble in water and therefore cannot be transported in solution in the plasma. Free fatty acids are transported bound to albumin, but the other lipids circulate in the plasma in lipoproteins, which are complex particles containing lipids and one or more of various apolipoproteins. Some apolipoproteins have a structural role; other functions include receptor recognition and the activation (and possibly inhibition) of enzymes involved in lipid metabolism. The function of the major apolipoproteins, which have an alphabetical classification and are abbreviated apo A, apo B, etc., are indicated in Table 5.1.

Lipoproteins have a non-polar core, containing cholesteryl ester and triacylglycerols, with an amphipathic surface layer of apolipoproteins and cholesterol, whose non-polar regions are directed inwards while the polar parts of the molecules are directed outwards (Figure 5.1).

Numerous disorders, both inherited and acquired, can affect the metabolism of lipids and lipoproteins, but the major clinical interest in them stems from the undoubtedly causal relationship between high concentrations of plasma lipids (particularly cholesterol) and the development of atherosclerosis, which is a disease of the arterial wall which leads to narrowing of the arterial lumen and a tendency to thrombus formation. When this affects the coronary arteries, it can cause angina pectoris, myocardial infarction or sudden death. It can also affect arteries supplying the brain (which can lead to transient ischaemic attacks and strokes), and cause peripheral vascular disease (manifested as claudication—sharp muscular pain on exercise—and gangrene).

Table 5.1 Classification and functions of apolipoproteins

Apoliprotein	Function
A-I	The major protein of HDL; activates lecithin:cholesterol acyltransferase
A-II	Structural in HDL; ?activates hepatic triglyceride lipase
A-IV	Unknown
B-48	Component of chylomicrons; cannot bind to LDL receptor
B-100 (a)	Component of LDL and VLDL; ligand for LDL receptor combines with apo B-100 in lipoprotein(a) which is an independent risk factor for coronary disease but whose function is unknown; structurally homologous with fibrinogen
C-I	Unknown
C-II	Present in chylomicrons and VLDL; activates lipoprotein lipase
C-III	Major component of VLDL; inhibits lipoprotein lipase
D	A minor apoprotein; function unknown
E	Present in all lipoproteins except LDL; ligand for remnant receptor and LDL receptor

This chapter discusses the normal metabolism of lipids and lipoproteins, the causes of abnormal lipid metabolism, the pathogenesis of atherosclerosis particularly with regard to lipids, approaches to the clinical management of lipid disorders, and the evidence that such intervention is beneficial.

5.2 The classification of lipoproteins

Lipoproteins have been traditionally classified on the basis of their density as determined by ultracentrifugation. Although this classification preceded a detailed appreciation of their metabolism, the classification remains useful, as each class of lipoprotein has a specific function. However, two important riders must be added to this statement. First, there is significant heterogeneity within each class of lipoprotein; i.e. the composition of the particles comprising each class is variable. Second, there is a dynamic relationship between the various lipoproteins, such that there is continuous exchange of components between them; furthermore, metabolism of lipoprotein particles can result in their being transformed to particles of a different class.

Figure 5.1

Cross-sectional diagram of a lipoprotein particle. The core consists of non-polar cholesteryl esters (CE) and triglycerides (TG), and the surface layer of proteins (PROT), phospholipids (PL) and cholesterol (CHOL), each with their polar domains exposed on the surface and their non-polar domains directed towards the non-polar core.

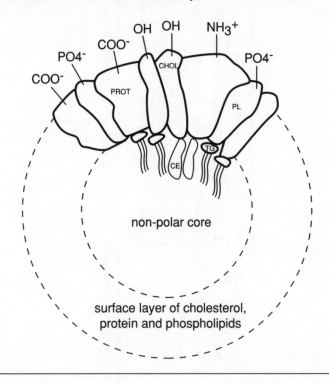

The major lipoproteins are chylomicrons (the least dense), very low-density lipoproteins (VLDL), low-density lipoproteins (LDL) and high-density lipoproteins (HDL) (Table 5.2). There are distinct sub-types of both LDL and HDL. Two other types are normally present in the blood in only very small amounts, but can accumulate in certain diseases. These are chylomicron remnants (CR) and intermediate-density lipoproteins (IDL). A further naturally occurring lipoprotein, lipoprotein(a), is normally present in only small, albeit highly variable amounts; its significance is discussed later. Finally, in some pathological states, other lipoproteins may be found (e.g. lipoprotein X in patients with cholestatic liver disease).

5.3 Lipoprotein metabolism

There are two main pathways of lipoprotein metabolism: the extrinsic pathway, responsible for the transport of dietary fat,

Table 5.2 Classes of lipoprotein

Particle	Major lipid	Major apoprotein	Density (g l^{-1})	Function
Chylomicrons	Triglyceride	B-48, C, E	<0.95	Transport of exogenous triglyceride
Very low-density lipoprotein (VLDL)	Triglyceride	B-100, C, E	0.96–1.006	Transport of endogenous triglyceride
Intermediate-density lipoprotein (IDL)	Cholesterol	B-100, E	1.007–1.019	Precursor of LDL
Low-density lipoprotein (LDL)	Cholesterol	B-100	1.02–1.063	Cholesterol transport
High-density lipoprotein (HDL)	Cholesterol, phospholipid	A, C, E	1.064–1.21	Reverse cholesterol transport

IDL is normally present in the blood in only small amounts. Note that subtypes of both LDL and HDL are recognised but that there is heterogeneity within all classes of lipoproteins.

and the intrinsic pathway, responsible for the transport of fat mainly arising by de novo synthesis. A third pathway, the 'reverse cholesterol transport' pathway, involving HDL cholesterol, interacts with both these.

In essence, dietary triglycerides (and other fats) are packaged into chylomicrons in the gut. Most of the triglyceride is taken up into adipose tissue and muscle, leaving remnant particles which are taken up by the liver. The liver synthesises VLDL; uptake of triglyceride into adipose tissue and skeletal muscle converts VLDL to IDL and these are either taken up by the liver or modified to form LDL, which transports cholesterol to peripheral tissues. These processes, and the involvement of HDL, will now be described in more detail.

5.3.1 Exogenous pathway

Dietary triglycerides are hydrolysed in the lumen of the gut by the action of pancreatic lipase to form free fatty acids and 2-monoglycerides. These are incorporated into micelles stabilized by bile salts and also containing other dietary lipids (e.g. cholesterol and fat-soluble vitamins). At the brush border of jejunal enterocytes, the lipids diffuse out of the micelles into enterocytes where the free hydroxyl groups of the monoglycerides are re-esterified with fatty acids to form triglycerides. These become the major components of chylomicrons (minor components include cholesterol, fat-soluble vitamins, apoprotein B-48 and various subtypes of apo A), which are secreted into the lacteals (intestinal lymphatic vessels) and reach the bloodstream through the thoracic duct. Apo B-48 is essential to the formation and transport of chylomicrons; each particle contains one molecule of this protein which remains associated with it throughout its life (fatty acids of 10–12 carbon atoms or less are not incorporated into triglycerides and diffuse into the blood where they are transported bound to albumin).

In the bloodstream, chylomicrons acquire other apolipoproteins, including apo C-II and apo E, phospholipids and cholesterol from HDL (see below). As the particles pass through capillaries in tissues such as adipose tissue and skeletal muscle, triglyceride is hydrolysed by an enzyme, lipoprotein lipase, associated with the capillary endothelium. This enzyme is activated by apo C-II. Free fatty acids thereby released are taken up into tissue cells to be used as sources of energy or, particularly in adipose tissue, to be re-esterified to form triglyceride. The glycerol released by the hydrolysis of triglycerides remains in the blood and eventually reaches the liver, where it is a substrate for glucose synthesis. As triglycerides are removed from chylomicrons, the particles become smaller and their surface area decreases. Surface components, including phospholipids, cholesterol and apo C-II are transferred back to HDL. As the particles shrink, their affinity for lipoprotein lipase decreases and, now classified as chylomicron remnants, they are transported to the liver. During this process, further modification occurs. Triglyceride is transferred to HDL in exchange for cholesteryl ester. The remnant particles, thus enriched in cholesteryl ester, are remove by the liver through interaction of apo E with a cell surface receptor, variously called the remnant receptor or LDL receptor related protein. The possible fates of cholesteryl ester in the liver include excretion in bile, conversion to bile acids and incorporation into VLDL (see below). This process is summarised in Figure 5.2.

Figure 5.2
The pathway of exogenous lipid metabolism. Dietary fat is absorbed into enterocytes and assembled into chylomicrons. The triglyceride in these particles is hydrolysed by lipoprotein lipase associated with the endothelium of capillaries in adipose tissue and skeletal muscle, releasing free fatty acids which are used as an energy substrate by muscle, but in adipose tissue are re-esterified to form triglycerides. The triglyceride-depleted remnant particles are removed from the circulation by the liver. Transfers of apolipoproteins are omitted for clarity: see text for details.

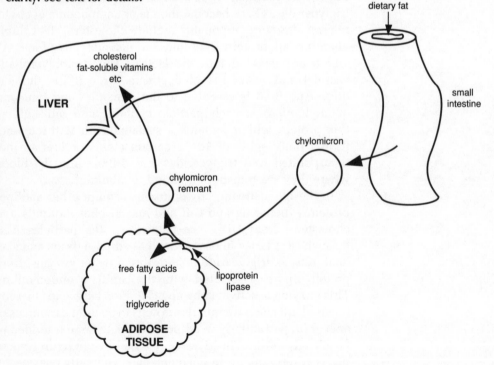

5.3.2 *Endogenous pathway*

The endogenous pathway shares some features with the exogenous pathway. The liver secretes triglyceride-rich VLDL particles into the bloodstream. The triglyceride can be derived from fatty acids taken up from the plasma (in turn derived from adipose tissue triglyceride) and from de novo synthesis. Each VLDL particle contains one molecule of apo B-100. This apolipoprotein is derived from the same gene as apo B-48, but, in the intestine, modification of the m-RNA results in premature termination of translation. Unlike apo B-100, apo B-48 is not capable of binding to the LDL receptor (see below).

As secreted, VLDL also contains some cholesterol, phospholipids and other apolipoproteins (apo E and apo Cs) but also acquire apolipoproteins from HDL in the bloodstream. VLDL particles show considerable variation in size; they are larger and are richer in triglycerides in obesity and diabetes mellitus and when there is excessive alcohol intake.

The initial fate of VLDL parallels that of chylomicrons, with lipoprotein lipase-mediated removal of triglyceride and transfer of surface components to HDL. The fate of the resulting remnant particles (IDL), however, is different from that of chylomicron remnants.

IDL can either be taken up by the liver, principally by interaction with the LDL receptor (also known as the apo B/E receptor, see below), or can be converted into LDL. This latter process involves further removal of triglyceride by hepatic triglyceride lipase, an enzyme associated with the endothelium of hepatic sinusoids.

It is of interest that, while both VLDL and IDL contain apo B-100, the principal ligand for the LDL receptor, it appears not to be accessible in VLDL, so that this lipoprotein cannot bind to the LDL receptor, and the binding of IDL is facilitated through apo E.

LDL is the principal cholesterol-carrying lipoprotein. Each particle contains a single molecule of apo B-100 (and very little other protein). When this protein binds to LDL receptors on the plasma membranes of hepatocytes and other cells surfaces, the receptor–lipoprotein complex is internalised by endocytosis. The receptor is recycled to the cell surface but the endosomes containing the lipoprotein particles fuse with lysosomes and free cholesterol is released into the cells.

This cholesterol may then be utilised in the synthesis of plasma and organelle membranes, undergo metabolic transformation (e.g. to steroid hormones in the adrenal cortex and gonads) or undergo esterification in a reaction catalysed by acylcoenzyme A:cholesteryl acyltransferase. Free (but not esterified) cholesterol inhibits the rate-limiting enzyme of cholesterol synthesis, hydroxymethylglutaryl-coenzyme A reductase (HMG-CoA reductase) and synthesis of LDL receptors. Thus an increased cellular demand for cholesterol would lead to both increased uptake from plasma LDL and increased intracellular synthesis. A decrease would have the opposite effect. This process is summarised in Figure 5.3.

Figure 5.3
The pathways of endogenous lipid metabolism. Very low-density lipoprotein (VLDL), containing mainly triglyceride synthesised in the liver and some cholesterol, is secreted from the liver into the blood. Much of its triglyceride is hydrolysed by lipoprotein lipase in adipose tissue and skeletal muscle, releasing free fatty acids into those tissues. The resulting intermediate-density lipoprotein (IDL) is then converted by hepatic triglyceride lipase to the cholesterol-rich low-density lipoprotein (LDL) which acts as a source of cholesterol for peripheral tissues. LDL is removed from the circulation by these tissues and by the liver by a receptor-mediated mechanism. LDL combines with LDL receptors clustered in coated pits and is internalised. Lysosomal enzymes release free cholesterol, and the LDL receptors are recycled to the cell surface. Free cholesterol inhibits intracellular cholesterol synthesis and LDL receptor synthesis but promotes cholesterol esterification. Cholesterol is the precursor of bile salts and is excreted from the liver in bile with bile salts as free cholesterol and cholesteryl esters.

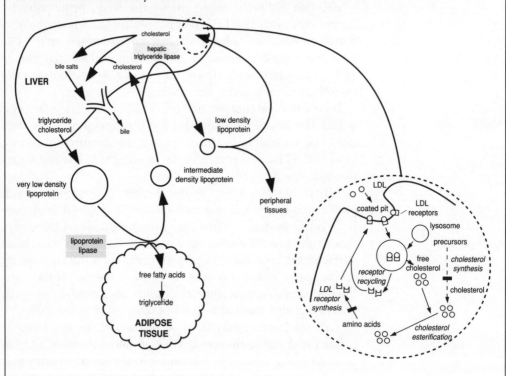

5.3.3 *HDL metabolism*

HDL has two principal functions: it supplies apolipoproteins to other lipoproteins, and is involved in the transport of cholesterol from peripheral tissues to the liver both directly and, indirectly, in remnant particles.

Its metabolism is complex. It is derived in precursor form ('nascent HDL') from intestinal cells and hepatocytes. Nascent HDL particles are disc-shaped and contain principally apoproteins (A-I and A-II) and phospholipids. They attract free cholesterol from tissues which is converted into cholesteryl ester by the action of lecithin:cholesterol acyl-transferase. This enzyme is activated by apo-AI. Cholesteryl ester is internalized and the particles become spherical HDL3 particles; as their volume increases, so does their surface area and they acquire further cholesterol, becoming less dense HDL2. HDL2 has several fates. It can be converted back to HDL3 by acquiring triglyceride from and losing cholesteryl ester to remnant particles, a process catalysed by cholesteryl ester transfer protein; it is probably also capable of being removed from the circulation by the liver. As has been indicated above, HDL also acts as a reservoir for certain apoproteins which are transferred to and accepted from other lipoproteins. This process is summarised in Figure 5.4.

5.4 Lipoprotein disorders

The most frequently occurring lipoprotein disorders are associated with increased plasma concentrations of cholesterol, triglycerides, or both. There are many such disorders, but only selected ones, having a particularly interesting molecular basis, are discussed in detail here. A comprehensive classification is given in Table 5.3 and readers seeking more information should consult the 'Further reading' section at the end of this chapter.

The clinical relevance of most, but not all, of the hyperlipidaemias relates principally to an increased risk of atherosclerosis (especially involving the coronary arteries) associated particularly with hypercholesterolaemia. Notable exceptions are conditions in which HDL concentrations are elevated, some of which are associated with a decreased risk of vascular disease, and isolated hypertriglyceridaemia, which carries a risk of pancreatitis but not of vascular disease.

Figure 5.4

A simplified diagram of reverse cholesterol transport by high-density lipoprotein (HDL). HDL is synthesised in a discoid form ('nascent HDL') in the liver and small intestine. It acquires surplus cholesterol from cell membranes and is converted into mature HDL. The cholesterol is esterified by lecithin:cholesterol acyl transferase to form cholesteryl ester (CE). Some HDL is probably catabolized by the liver but much cholesteryl ester is transferred to remnant particles (chylomicron remnants and intermediate-density lipoproteins) by cholesteryl ester transfer protein (CETP) and reaches the liver in these particles. Note that HDL is also involved in the transport of apolipoproteins between lipoproteins and that CETP transfers triglyceride from remnant particles to HDL (converting HDL3 particles into HDL2) but these details are omitted for the sake of clarity.

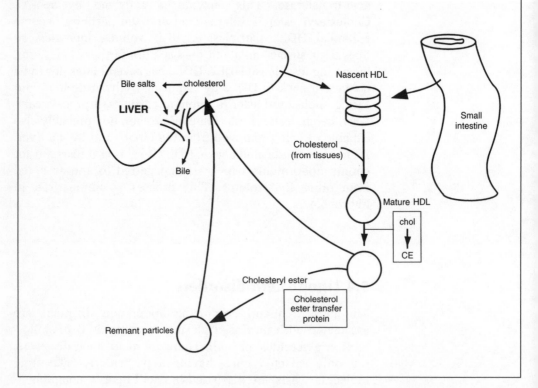

Disorders characterised by low concentrations of lipoproteins also occur, but are rare. They are not considered further in this chapter.

5.4.1   *What is hyperlipidaemia?*

It has been suggested that patients' alcohol intakes should be considered excessive when they exceed those of their doctors. Plasma cholesterol concentrations are sometimes considered in the same light. When a measurement of some natural variable is made in an individual, it has little mean-

Table 5.3 The major inherited hyperlipoproteinaemias

Type	Particle present in excess	Lipid abnormality		Molecular basis
		Cholesterol	Triglyceride	
Familial hypercholesterolaemia	LDL	↑↑	N	Defective synthesis/function of LDL receptor
Polygenic hypercholesterolaemia	LDL	↑	N	Unknown
Remnant dyslipidaemia	IDL	↑	↑	Abnormal apo E
Familial combined hyperlipidaemia	VLDL	↑	↑	Increased apo B synthesis
Chylomicronaemia syndromes	CM	N	↑↑	Lipoprotein lipase or apo C-II deficiency
Familial hypertriglyceridaemia	VLDL, (CM)	N	↑↑↑	Unknown

Note that the severity of the hyperlipidaemia may depend on environmental factors in addition, especially in polygenic hypercholesterol-aemia, remnant dyslipidaemia, familial combined hyperlipidaemia and familial hypertriglyceridaemia.
N = normal.

ing until it is compared with a standard. Conventionally, the comparator is the 'normal range'—i.e. the range of values into which the same measurement would fall in 95% (the range from two standard deviations below the mean to two above) of healthy individuals. While this is appropriate for many measurements, it is misleading for some, where what is apparently normal may not necessarily be devoid of risk. Measurements of serum cholesterol concentration are a case in point. There is no doubt that there is a continuous relationship between plasma cholesterol concentration and coronary heart disease mortality, even at supposedly low plasma concentrations. (There is an increase in overall mortality associated with very low plasma cholesterol concentrations, but this appears to be a consequence of disease on cholesterol rather than vice versa.) Thus concentrations found in healthy people may be associated with a measurable (albeit usually small risk) of coronary disease. This is illustrated in Figure 5.5. The 'normal range' for cholesterol in adult males with no clinical evidence of coronary disease is approximately 3.6–6.7 mmol l^{-1}, yet the coronary risk associated with a cholesterol of 6.7 mmol l^{-1} is approximately twice that associated with one of 5.2 mmol l^{-1}.

Thus while the term 'hypercholesterolaemia' is often applied when the total serum cholesterol concentration exceeds 6.7 mmol l^{-1}, concentrations lower than this should not necessarily be accepted as 'normal', in the sense of 'desirable' or 'free of risk', particularly in individuals whose risk of coronary disease is increased for other reasons (see below). In considering individuals, it is better to consider what concentration of cholesterol is desirable, rather than how that individual's cholesterol compares with other people's.

The same problem does not arise with triglyceride, for which a fasting concentration of <1.8 mmol l^{-1} is usually taken as 'normal'; there is little evidence of any gradation of risk at lower concentrations.

5.4.2 Classification of hyperlipidaemias

Hyperlipidaemias are classified as primary (i.e. genetically determined), or secondary (i.e. related to the presence of some other disorder or treatment with certain drugs). The distinction is important: if a hyperlipidaemia is secondary to another condition, treatment should be directed towards that condition. On the other hand, patients with primary hyperlipidaemias may require specific intervention, often with diet, and sometimes also with drugs. As will be dis-

Figure 5.5
Plasma cholesterol concentrations and risk of coronary heart disease. The risk of coronary disease increases with increasing cholesterol concentration, but this occurs even within the range of concentrations found in apparently healthy people.

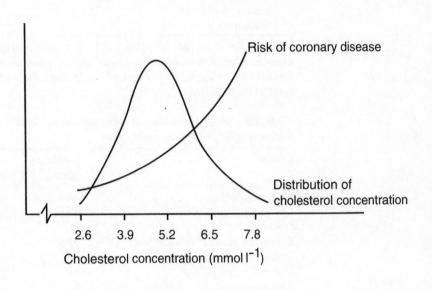

cussed in detail later, however, hyperlipidaemias should not be managed in isolation. Other risk factors for vascular disease (see Table 5.4) must also be assessed and, if possible, treated.

5.4.2.1 *Secondary hyperlipidaemias*

Some of the more frequent secondary causes of hyperlipidaemia are listed in Table 5.5. Hypothyroidism (which causes a decrease in the synthesis of LDL receptors) is of particular importance because it is common, particularly in older people, and because its onset is frequently insidious. The incidence of hypothyroidism in patients investigated for hypercholesterolaemia has been reported to be as high as 5%. Excessive alcohol intake and simple obesity are the most frequent causes of hypertriglyceridaemia. Non-insulin-dependent diabetes mellitus is another common cause of hypertriglyceridaemia, and the insulin resistance frequently present in this condition is also thought to be responsible

Table 5.4   Major risk factors for coronary heart disease (CHD)

Potentially modifiable	Unmodifiable
• Cigarette smoking • Hypertension • Hypercholesterolaemia • Hypertriglyceridaemia + low HDL cholesterol • Diabetes mellitus* • Inadequate exercise	• Male sex • Personal history of CHD • Family history of premature CHD

*The excess CHD risk associated with non-insulin-dependent diabetes is related to insulin resistance, and may not be decreased by improved glycaemic control unless this is associated with improved sensitivity to insulin.

Table 5.5   Secondary causes of hyperlipidaemias: effects of other conditions on plasma lipid concentrations

Cause	Increased cholesterol	Increased triglyceride
Hypothyroidism	+++	+
Cholestatic liver disease	+++	N
Obesity	N	+
Excessive alcohol consumption	N/+*	++
Diabetes mellitus**	N	++
Chronic renal failure	N/+	+
Thiazide diuretics	N	+
Beta-adrenergic antagonists (some)	+	+

*HDL cholesterol is increased by moderate alcohol consumption, and also by exercise.
**In well-controlled insulin-dependent diabetes, lipids may be normal (HDL can be elevated), but in non-insulin-dependent diabetes the abnormalities shown, together with a low HDL, frequently persist even if glycaemic control is improved.
N = normal.

for the low concentrations of HDL cholesterol and for the tendency for LDL particles to be both more dense and triglyceride-rich (and atherogenic) in this condition.

Some drugs can also cause, or exacerbate pre-existing, hyperlipidaemia. The list includes some that are widely used in the management of cardiovascular disease (e.g. certain beta-adrenergic antagonists and thiazide diuretics).

5.4.2.2 *Primary (genetic) hyperlipidaemias*

FAMILIAL HYPERCHOLESTEROLAEMIA (FH)

This condition is probably the most fully studied and best understood hyperlipidaemia. It is inherited as an autosomal dominant condition, being manifest in heterozygotes (see Chapter 1). In the UK, the approximate incidence is 1 in 500, and 50% of people with this condition, if untreated, will have clinical evidence of coronary heart disease by the age of 50 years. The incidence of homozygosity is approximately 1 in 10^6. Homozygotes (in fact, because many mutations exist, most are compound heterozygotes) develop coronary disease in their teenage years.

Pathogenesis   The defect is a decrease in the number of functional LDL receptors; heterozygotes have approximately half the normal number while homozygotes effectively have none. Because of decreased uptake of LDL, and increased synthesis from IDL whose uptake is also dependent on the LDL receptor, LDL cholesterol concentrations in the plasma are greatly increased.

Five classes of mutation have been described in FH.

- Class 1 (null mutations), the commonest, are associated with an absence of receptor synthesis; the other classes are associated with normal synthesis.

- Class 2 (transport-deficient mutations) cause impaired transport of receptors from the smooth endoplasmic reticulum to the Golgi apparatus.

- Class 3 (binding-deficient mutations) cause the formation of a receptor which cannot bind the ligand.

- Class 4 (internalisation-defective mutations) result in the formation of a receptor which is unable to cluster in coated pits on cell surfaces.

- Class 5 (recycling-defective mutations) are associated with a failure of the normal recycling of internalised receptors to the cell surface.

Diagnosis   Patients with FH may present with coronary disease or with xanthomata—subcutaneous accretions of cholesterol which in FH characteristically affect certain tendon sheaths—but an increasing number are being detected as a result of screening programmes. It has been advocated that the whole adult male population (coronary heart disease is uncommon in premenopausal women; after the menopause, the risk increases to become similar to that of males; this point is discussed further in a later section of this chapter) should be screened for hypercholesterolaemia but this would be a mammoth task and at present much screening is opportunistic (taking place when an opportunity presents itself, for example, during the course of a consultation with a doctor for some other reason, or when an individual uses a self-testing kit or has a test performed in a high-street setting), or is directed towards people at high risk (e.g. those with family histories of premature coronary disease or who are members of families in which a case of FH has been identified).

Familial hypercholesterolaemia is diagnosed on the basis of a serum cholesterol concentration >7.8 mmol l^{-1} in adults, in the absence of secondary causes of hyperlipidaemia and in association with evidence of hypercholesterolaemia in first-degree relatives. Some patients with FH have tendon xanthomata and these are considered to be pathognomonic of the condition, but they are often not present, particularly in younger patients. A family history of hypercholesterolaemia may not be available, but the diagnosis should be suspected if there is a family history of premature coronary disease (i.e. occurring before the age of 60 years). The number of mutations that can cause the disorder makes it impractical to attempt routine genetic diagnosis and it is not necessary for effective treatment.

Management The management of genetic hyperlipidaemia should always start with diet. The general principles are to reduce dietary fat intake so that it provides not more than 30% of total energy, to reduce saturated fat to not more than 33% of the total, and to reduce overall energy intake in order to achieve a body weight that is appropriate for height. Reduction of the ratio of waist and hip circumferences (a marker of central adiposity, an independent coronary disease risk factor) may also be desirable.

In the majority of patients with FH, however, diet alone will be insufficient to lower the cholesterol to desirable levels and drug treatment will be required in addition.

The drugs of choice in the majority of patients are a class known as statins. These are inhibitors of the rate-limiting enzyme of cholesterol synthesis, HMG-CoA reductase (see above). Inhibition of this enzyme decreases cholesterol synthesis and thus intracellular cholesterol concentrations, thus decreasing the inhibition of LDL receptor synthesis by cholesterol (see Figure 5.3). The increase in LDL receptor number increases receptor-mediated LDL uptake from the blood and lowers its concentration, thus decreasing the amount available for uptake by macrophage scavenger receptors in the arterial wall, a key event in atherogenesis (see below).

The other major class of cholesterol-lowering drugs are bile acid sequestrants. Some 90% of bile acids secreted into the gut are normally reabsorbed and returned to the liver (enterohepatic circulation). Bile acid sequestrants bind bile acids in the gut and interrupt this process. Hepatic bile acid concentrations fall, stimulating bile acid synthesis from cholesterol, and thus depleting hepatic cholesterol and reducing the synthesis of LDL. Although there are drawbacks to their use (they may cause unpleasant gastrointestinal side-effects), bile acid sequestrants are powerful cholesterol-lowering agents. Because their mode of action is distinct from that of the statins, the effects of these agents used in combination is summative, and used together they can provide effective treatment for patients with severe hypercholesterolaemia whose cholesterol remains higher than desirable on treatment with a single agent alone. Patients with homozygous FH do not respond to drug treatment. Techniques such as LDL apheresis, involving regular physical removal of LDL particles from the blood, have been used with success in such patients, and this technique is also effective in heterozygotes in whom treatment with diet and drugs does not lower the cholesterol sufficiently. As mentioned in Chapter 7, liver transplantation has also been used in homozygous FH.

FAMILIAL DYSBETALIPOPROTEINAEMIA (ALSO KNOWN AS REMNANT HYPERLIPIDAEMIA AND BROAD BETA DISEASE)

This is a rare disorder, occurring in approximately 1 in 10 000 people in the UK. It is characterised by the accumulation in the plasma of IDL (intermediate-density lipoproteins) and chylomicron remnants; this causes an increase in cholesterol and triglyceride concentrations and patients may have characteristic types of xanthomata, affecting the elbows, knees, trunk and the palms of the hands. This condition is associated with an increased risk not only of cardiovascular disease, as might be expected, but also of peripheral vascular disease, which marks it out from the other genetic dyslipidaemias. The old name for the condition, 'broad beta disease', stems from the appearance of serum lipoproteins when subjected to zone electrophoresis on, for example, agarose gel. The presence of high levels of IDL causes a characteristic broad band between the areas normally occupied by LDL and VLDL (corresponding to the electrophoretic mobility of β-globulins).

Pathogenesis   The molecular basis of familial dysbetalipo-proteinaemia is well understood: the gene for apolipoprotein E exhibits polymorphism, giving rise to similar, but distinct gene products. The most frequent genotype is designated *e3/e3* and gives rise to the phenotype E3/E3. Individuals who are homozygous for the *e2* gene (*e2/e2*) express an abnormal apoplipoprotein E (E2), which has reduced capacity for binding to apo B/E (LDL) receptors.

What is particularly interesting is that the *e2/e2* genotype occurs in approximately 1 in 100 individuals whereas remnant disease is 100 times rarer. That is, the majority of individuals with the abnormal genotype do not develop the clinical disorder. This is an example of the importance of the interaction between genetic and environmental factors in the pathogenesis of disease. It appears that, in order for the abnormal genotype to be expressed phenotypically, other factors need to be present. These include being overweight, having diabetes mellitus and having hypothyroidism. Each of these conditions itself can cause lipid abnormalities in normal individuals, but it seems that they are more likely to do so against the background of a genetically determined abnormality in lipid metabolism.

Management   The lipid abnormalities in patients with this condition usually respond readily to treatment. Precipitating conditions should be identified and treated if present; dietary advice should be aimed at normalising body weight and reducing fat intake. Some patients will need lipid-lowering drugs in addition, usually of the fibrate class, which act principally by reducing the synthesis of VLDL.

CHYLOMICRONAEMIA SYNDROMES

Rarer still are conditions in which chylomicrons are present in the fasting state. They are, as has been discussed above, the major carriers in the blood of dietary triglycerides, but are usually rapidly cleared. Fasting chylomicronaemia, associated with high plasma triglyceride concentrations, is characteristic of these conditions, which are associated with a risk of acute pancreatitis but not of vascular disease. This is thought to be related to obstruction of pancreatic capillaries by excess chylomicrons and consequent inflammation of the pancreas itself.

The condition usually becomes apparent in infancy, because of episodes of abdominal pain, the presence of the characteristic eruptive xanthomata, or because a blood sample taken for some other reason is noticed to have a milky appearance ('lipaemia').

Pathogenesis Two rare inherited disorders are associated with the chylomicronaemia syndrome. The inheritance of both is autosomal recessive. Both are predictable on the basis of our understanding of the metabolism of chylomicrons. These lipoproteins are metabolised through interaction with the enzyme lipoprotein lipase, which requires apoprotein C-II (a component of chylomicrons) for its activation. Defective synthesis of either the enzyme or apo C-II can lead to decreased metabolism of chylomicrons in the blood and consequent hypertriglyceridaemia. In general, the effects of apo C-II are less severe than lipoprotein lipase deficiency, possibly because some metabolism of chylomicrons by lipoprotein lipase does occur when they are present at high concentrations even in the absence of the enzyme's activator.

Diagnosis Lipoprotein lipase deficiency can be detected by attempting to measure the enzyme in the plasma following administration of heparin (which normally causes some release of the enzyme from the endothelial cells to which it is bound). Apo C-II deficiency can be diagnosed using immunoassay techniques.

Management The mainstay of treatment of chylomicronaemia syndromes is dietary. The diet should be generally low in fat; the use of triglycerides containing medium-chain fatty acid residues is helpful as these are absorbed from the gut directly into the bloodstream, without being packaged into chylomicrons.

OTHER GENETIC CAUSES OF HYPERLIPIDAEMIA

Table 5.3 indicated that there are many other causes of hyperlipidaemia. These include familial combined hyperlipidaemia, which occurs in approximately 1 in 100 people and is due to increased synthesis of apo B-100 (the reason for this is not known), and can lead to hypercholesterolaemia, hypertriglyceridaemia, or both.

In many individuals with hypercholesterolaemia, other members of the family are affected, but there is no clear pattern of inheritance. The term common or polygenic hypercholesterolaemia is often used to describe this condition. Both these terms are appropriate but the condition is not precisely defined. Cholesterol concentrations are lower than in FH, and appear to be more affected by environmental factors, such as diet and body weight.

5.5 The concept of coronary heart disease risk and the appropriate use of lipid-lowering drugs

The rationale for attempting to lower a patient's plasma lipid concentrations if they are abnormally high is based on the premise that doing so will reduce the risk of the consequences of hyperlipidaemia (i.e. in most cases, the risk of coronary heart disease). In patients with very high triglyceride concentrations, the aim of treatment is to reduce the risk of pancreatitis.

There are four important questions to consider. These are:

1. Is there any evidence that lowering plasma lipid concentrations is actually beneficial? This question can be answered in the affirmative for certain groups of patients and is discussed in more detail in the next section.

2. Are there any potential disadvantages of lowering plasma lipid concentrations, or adverse effects of the treatments used to achieve this? Unfortunately, no drugs are free of a risk of causing adverse effects in some individuals, although this is fortunately uncommon with fibrates and statins, two of the major classes of lipid-lowering drugs. Whether lowering cholesterol itself can be harmful is discussed below.

3. At what lipid concentrations should we intervene, and what should be the aims of treatment?

4. Are there other risk factors that need to be taken into account?

The last two questions are closely related. Although hypercholesterolaemia (and, to a lesser extent, hypertriglyceridaemia) are risk factors for coronary heart disease, there are many others (see Table 5.4). Of the most important, some are potentially modifiable (e.g. hypertension and cigarette smoking) whereas others are not (e.g. male sex, increasing age and a family history of coronary disease occurring before the age of 60 years). Most importantly, individuals with a personal history of coronary disease (e.g. angina or a previous myocardial infarct) are at high risk of having a (further) myocardial infarct. All these factors need to be taken into account in assessing whether to institute lipid-lowering treatment: a plasma cholesterol concentration of, say, $6.5\,\mathrm{mmol\,l^{-1}}$ poses little risk in a man of 40 years who has no other risk factors, but would be unacceptably high in someone with multiple risk factors and particularly in someone who had had a previous myocardial infarct.

It is not appropriate to discuss here the finer points of treatment of the hyperlipidaemias, but the general principles are straightforward. Secondary causes should be sought and treated appropriately if present; the first-line treatment should also be diet (and other lifestyle measures, such as increasing aerobic exercise, which may aid weight loss, improve cardiovascular fitness and increase HDL cholesterol). But consensus guidelines now suggest that, if an adequate trial of simple measures alone does not lower the cholesterol below the limits shown in Table 5.6, pharmacological intervention should be instituted.

Guidelines are, however, no more than their name suggests, and informed physicians will always make decisions based on their individual judgement, taking into account such factors as the relative concentrations of LDL and HDL cholesterol, the extent of any hypertriglyceridaemia, the patient's general health and, of course, the patient's wishes.

It might be suggested that, since there is a continuing relationship between coronary risk and cholesterol concentration even at relatively low concentrations, cholesterol-lowering would be advisable for everyone, not just people with high concentrations or having other coronary risk fac-

Table 5.6 Target values for serum cholesterol concentration

Category	Cholesterol concentration $(\mathrm{mmol\,l^{-1}})$	
	Total	LDL
Pre-existing coronary disease revascularisation procedure, heart transplant	<5.2	<3.25
Multiple risk factors or genetically determined hyperlipidaemia (e.g. familial hypercholesterolaemia)	<6.5	<5.0
Asymptomatic hypercholesterolaemia with no other risk factors	<7.8	<6.0

If levels greater than the target values persist in spite of optimisation of diet, treatment with lipid-lowering drugs should be considered. Note that different levels may be applicable in different patients; guidelines or targets should not be used indiscriminantly without a complete clinical assessment.

tors, particularly since there is no clear evidence of a harmful effect of lowering cholesterol or of a low cholesterol. This philosophy underlies the public health approach to the decrease in coronary risk through the advocacy of changes in diet and lifestyle for everyone. But lipid-lowering drugs, although generally safe, can have adverse effects and have to be paid for. Although the evidence from clinical trials supports the idea of intervention with lipid-lowering drugs along the lines suggested above, the financial burdens of such a policy would be considerable.

5.5.1 *Evidence of benefit from lipid-lowering*

Although the importance of hypercholesterolaemia in the pathogenesis of coronary artery disease has been appreciated for many years, it is only relatively recently that unequivocal evidence of benefit from lipid-lowering has been forthcoming. It would be beyond the scope of this book to discuss this evidence in detail, and the interested reader should consult references cited in 'Further reading' at the end of this chapter, but broadly speaking, it falls into two categories. Regression studies have been aimed at quantifying the effect of lipid-lowering on the narrowing of coronary arteries by atherosclerosis, using serial angiography to measure the diameters of these vessels. Such studies have generally shown that lowering cholesterol is associated in significant numbers of cases (though by no means all) with either a decrease in the rate of progression of disease with time or actual regression of lesions (i.e. an increase in luminal diameter and thus the potential for blood flow).

More importantly, well-conducted clinical trials have shown a decrease both in coronary events and in overall mortality in association with lowering cholesterol. This has been achieved in the context of primary prevention (i.e. in individuals with no clinical evidence of coronary disease on entry to the trial) and secondary prevention (in individuals with a history of angina or previous myocardial infarction). The most powerful secondary prevention study (the 4S study), conducted in Scandinavia and involving 4444 individuals, demonstrated that treatment with simvastatin lowered total cholesterol concentrations by a mean of 25% (and LDL cholesterol by 35%), reduced the overall mortality in the treated group by 30% and the risk of death from coronary disease by 42%. As with many previous studies, most of the subjects were men and below the age of 60 years, but there were sufficient over the age of 60 years to show benefit in this age-group too; although the results for women did not achieve statistical significance, there was a trend towards decreased mortality and coronary events in women also. Of particular interest is the fact that the range of total cholesterol concentrations on entry to the study was 5.5–8.0 $mmol\,l^{-1}$ (mean 6.75), which is only moderately higher than ideal by conventional criteria; furthermore, the benefits of treatment appeared to be as great for participants whose cholesterol concentrations fell into the lowest quartile as those in the highest. There was no evidence for any detrimental effect of lowering cholesterol, with no increase in non-cardiac deaths (e.g. from cancer) in the treated group. This last observation is of especial importance, since this possibility had been suggested (and seized upon by the lay press) by the results of some earlier, smaller trials in which there were apparent (but non-significant) increases in non-cardiac mortality in treated groups. (It is now generally accepted that the only proven adverse effect of lowering cholesterol concentrations is a very small increase in the risk of haemorrhagic stroke.) It is also noteworthy that, although it was not a specific objective of the trial, the 4S study demonstrated a reduction in the number of patients sustaining cerebrovascular events (strokes). Atherosclerosis is the commonest cause of strokes, but the importance of cholesterol-lowering in patients with stroke is also related to the fact that approximately one-half of patients who survive a stroke will die of coronary, not cerebrovascular, disease.

Broadly similar results were obtained in a large study of primary prevention of coronary disease in men in the west of Scotland, an area with a high prevalence of this condition. Treatment with pravastatin reduced total cholesterol concentrations by an average of 20% (LDL by 26%), coronary events by 31% and overall mortality by 22%.

Many questions remain to be answered, even with secondary prevention. Is there a lower limit beyond which there is no benefit of lowering cholesterol further?* Can the results in men be extrapolated to women? Is the drug used for lowering cholesterol important? In relation to primary prevention, perhaps the most important question to resolve is, when is intervention with drugs appropriate? All these, and many other questions, will no doubt eventually be answered by the results of clinical trials. Meanwhile no-one would argue that patients with a personal history of vascular disease or familial hypercholesterolaemia should be treated vigorously, as should individuals with moderate hypercholesterolaemia but having other (and unmodifiable) risk

*See Note in proof, p. 106.

factors for coronary disease. But while it might be ideal to attempt to lower everyone's cholesterol to $5.2\,\mathrm{mmol\,l^{-1}}$ or some other 'ideal' level, this is clearly not practicable. The best approach is likely be to continue to encourage dietary habits and a lifestyle that may reduce the risk of coronary disease in the population at large, while continuing to identify and target those at higher risk for pharmacological intervention.

5.6 Atherogenesis

This condition, which is the major cause underlying cause of myocardial infarction and strokes, is a degenerative disease of the arterial wall characterised by lipid deposition and fibrosis. Despite an immense amount of research, some details of its pathogenesis remain unclear, but the stages of development have been well defined.

The initial lesion is the fatty streak. This involves the focal accumulation of cholesteryl ester from the plasma within the intima of the affected arteries. Exactly how this happens is uncertain. Hypercholesterolaemia itself may cause endothelial dysfunction and facilitate the ingress of LDL from the plasma; certainly endothelial damage as a result of smoking (another important risk factor) can accelerate the process. It has become clear that oxidation of LDL renders it more atherogenic than normal LDL (hence the current interest in the use of antioxidants such as vitamin E). Oxidation may occur in the plasma but certainly can take place in the intima. Oxidised LDL is chemotactic to blood monocytes which then engulf the LDL through non-saturable scavenger receptors to form fat-laden tissue macrophages (foam cells), the typical cells of the fatty streak.

The release of a wide range of chemotactic and growth factors (e.g. platelet-derived growth factor) by macrophages leads to further cellular infiltration and proliferation of smooth muscle cells which synthesise collagen, elastin and mucopolysaccharides, so leading to the formation of plaques with a fibrous cap overlying a lipid-rich core. Such lesions have a covering of vascular endothelium and appear to be relatively stable. Initially, these lesions are of little clinical significance, but progressive enlargement may lead to significant narrowing of the arterial lumen. More importantly (and dramatically), superficial injury to the endothelial surface can expose underlying collagen and lead to thrombus formation, while plaque rupture due to accumulation of excess lipid can lead to bleeding into the plaque itself, again precipitating thrombosis. The thrombus usually occludes the vessel, causing ischaemia of the muscle supplied by the artery and, unless perfusion can be restored by these of thrombolytic drugs (e.g. streptokinase) or mechanically (e.g. by angioplasty), myocardial infarction will result. The structure of an atheromatous plaque is illustrated diagramatically in Figure 5.6.

Figure 5.6
Cross-sectional diagram of an artery with an atheromatous plaque. Note the narrowing of the vessel lumen; rupture of the plaque cap exposes the thrombogenic material within the plaque to the blood, causing thrombus formation and often total vascular occlusion.

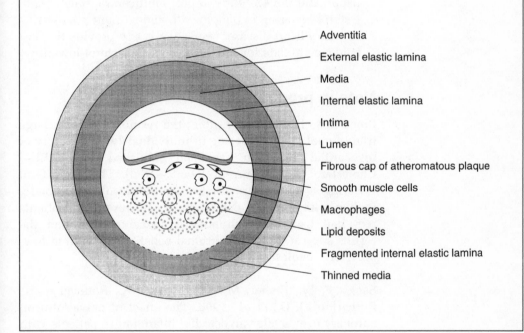

Adventitia

External elastic lamina

Media

Internal elastic lamina

Intima

Lumen

Fibrous cap of atheromatous plaque

Smooth muscle cells

Macrophages

Lipid deposits

Fragmented internal elastic lamina

Thinned media

Further reading

Feher, M. and Richmond, W., 1996. *Lipids and Lipid Disorders,* 2nd edn, London: Mosby-Wolfe. (A short, well-illustrated book summarising the physiology and pathology of lipid metabolism, clinical trials and treatment.)

Hunningshake, D.B. (Ed.), 1994. Lipid disorders. *Medical Clinics of North America,* **78,** 1–226. (A set of authoritative review articles covering all aspects of the topic.)

Law, M.R., Thompson, S.G. and Wald, N.J., 1994. Assessing possible hazards of reducing serum cholesterol. *British Medical Journal*, **308**, 373-379 (This and the two papers which immediately follow it—now colloquially known as 'the cholesterol papers'— summarise the evidence of benefit from cholesterol-lowering and refute many of the arguments advanced against it.)

Scandinavian Simvastatin Survival Study Group, 1994. Randomized trial of cholesterol lowering in 4444 patients with coronary heart disease: the Scandinavian Simvastatin Survival Study (4S). *Lancet* **344,** 1383–1389.

Shepherd, J., Cobbe, S.M., Ford, I., Isles, C.G. et al., 1995. Prevention of coronary heart disease in men with hypercholesterolaemia. *Lancet* **333,** 1301-1307. (This paper, and the 4S Study, report multicentre trials of cholesterol-lowering in people with and without pre-existing coronary heart disease, respectively, and provide the best evidence to date for the benefits of cholesterol-lowering.)

Note in proof

Since this chapter was written, the results of another large trial of cholesterol lowering in individuals with a history of myocardial infarction has been published (the CARE—Cholesterol and Recurrent Events—trial). The cholesterol lowering agent was pravastatin, and the results are broadly in line with those of the 4S study. However, the patients recruited had lower cholesterol concentrations, and the results suggest that no additional benefit accrues from lowering LDL cholesterol below $3.25 \, \text{mmol} \, \text{l}^{-1}$.

Sacks, F. M., Pfeffer, M. A., Moye, L. A., Rouleau, J. L., Rutherford, J. D., *et al.*, 1996. The effect of pravastatin on coronary events after myocardial infarction in patients with average cholesterol levels. *New England Journal of Medicine*, **335**, 1001–1009.

6 Molecular Histopathology

Jonathan R. Salisbury

6.1 Cell proliferation

Pathologists are interested in cell proliferation not only because it is one of the vital cellular mechanisms but also because it may give prognostic information about neoplasms. One of the prognostic assessments that can be made on a light-microscopic section of a neoplasm is that of the histological grade of the tumour (Box 6.1). Grade does predict biological behaviour quite well for some tumours (e.g. squamous carcinoma of the skin). Unfortunately, for many tumours there is a rather poor correlation between grade and prognosis. Some of this may be related to difficulties in assessing grade. If you examine a section of a neoplasm it is apparent that the differentiation of the tumour cells varies from area to area. The histopathological diagnosis is made on the best area(s) whilst the assessment of differentiation should be made on the worse area(s), but what if these are only a tiny fraction, perhaps 1%, of the whole? Is the worse area even on the section or is it still in the tissue block or the specimen pot? Adequate sampling is clearly vital. Recognising mitoses in light-microscopic sections can also be difficult. Van Diest and Baak (1992) give useful criteria for the identification of mitotic figures. The nuclear membrane must be absent (so the cell must have passed prophase). Clear, hairy extensions of nuclear material (condensed chromosomes) must be present,

Box 6.1 **Histological grade of a neoplasm**

- The grade is a combination of the cytological differentiation (i.e. maturity) and architecture of the tumour cells and the number of mitoses. The presumption is that the grade correlates with the biological behaviour of the tumour (i.e. the greater the grade, the more aggressive the tumour).
- Grade can be expressed numerically: e.g. Broder's grades for squamous carcinoma (grades 1–4) or the grades of chondrosarcoma (grades 1–3).
- Alternatively, grade can be expressed ordinally as low-grade, intermediate grade or high-grade: e.g. for non-Hodgkin's lymphomas.

either clotted (start of metaphase), in a plane (metaphase/anaphase), or in separate clots (telophase) (Figure 6.1).

Figure 6.1

The stages of mitosis. The separation of the duplicate copies of the genome has six recognisable phases by light microscopy.

Prophase

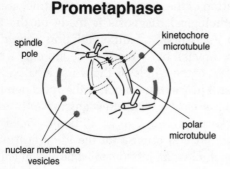

nuclear membrane

centriole centre of spindle

centromere

two sister chromosomes held together at centromere

microtubules of spindle

Anaphase

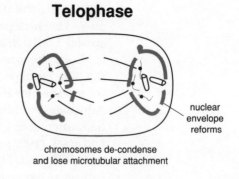

nuclear envelope vesicles migrate towards poles

elongation of polar microtubule

shortened kinetochore microtubule

chromatids pulled towards pole of spindle

Prometaphase

spindle pole

kinetochore microtubule

nuclear membrane vesicles

polar microtubule

Telophase

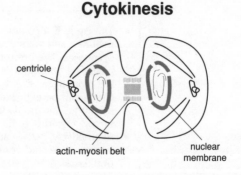

nuclear envelope reforms

chromosomes de-condense and lose microtubular attachment

Metaphase

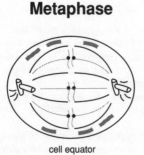

cell equator

Cytokinesis

centriole

actin-myosin belt

nuclear membrane

Condensed mitotic figures can simulate apoptotic bodies very well in tissue sections (Box 6.2). Proliferative activity is defined as the number of cells that can be produced by a cell population in a given time. Usually proliferative activity is assessed by calculating indexes such as the mitotic index, the S phase fraction, or the Ki67 labelling index (Box 6.3, p. 111). However, an increase in one of these indexes does not necessarily mean that there is an increase in cell production. The more the mitotic process is impaired, for example, then the greater the mitotic index. This is exemplified by colchicine arrest. For most tumours, these indexes do not really represent cell production and, therefore, do not represent proliferative activity.

There is also a clear difference between cell production and growth of a tumour. The latter is affected by apoptosis

Box 6.2 **Apoptosis 'programmed cell death'**

- Apoptosis is a controlled, biologically purposeful, process that deletes single cells whose continued viability would prejudice the functional integrity of the tissue. It is also called programmed cell death. Normal cells require the expression of certain genes to avoid apoptosis.
- The cells shrivel to produce apoptotic bodies which are then phagocytosed by neighbouring parenchymal cells or by macrophages in transit through the tissue. Apoptotic bodies appear as small membrane-bounded, densely eosinophilic cytoplasmic bodies with either a single pyknotic nucleus or several nuclear fragments.
- Apoptosis is seen in many tissues and is, for example, the mechanism of morphogenesis in embryos and of metamorphosis in larval forms (e.g. the tadpole losing its tail to be become a frog).

- In normal human tissues it can be seen particularly well in premenopausal endometrium, in haemopoietic bone marrow, in thymus, and in the germinal centres of lymphoid follicles where B cells that have been unsuccessful in their gene rearrangements (see p. 115) undergo apoptosis and are phagocytosed by macrophages to create 'tingible bodies' (tingible = stainable).
- In human neoplasms, apoptosis is abundant in basal cell carcinomas of the skin, hence their clinically slow growth despite numerous mitotic figures, and in Burkitt's lymphomas which have a 'starry sky' appearance at low power because of the numerous pale-staining apoptotic cell-containing macrophages scattered amongst the darkly-stained proliferating lymphoblasts.

Box 6.3 **Proliferative activity indexes**

- Mitotic index = growth fraction × mitotic phase/cell cycle time.
- S phase fraction = growth fraction × S phase/cell cycle time.
- Growth fraction = proportion of cells that are cycling in that tissue.

- Cycle time = sum of duration of individual phases.
- Growth fraction, M phase, S phase and cycle time are all biologically independent. They are intrinsically regulated by the cell and not by signals that the cell receives from the cell population.

(Box 6.2) and tumour necrosis because of ischaemia, and reflects the difference between cell gain and cell loss. Cell production is determined by the growth fraction and the cell cycle time.

Antibodies recognising many of the 'cell cycle proteins' are now widely used.

6.2 Cell cycle proteins

6.2.1 Cyclins

Cyclins are a family of regulatory subunits that act, together with a family of proteins called cyclin-dependent kinases (CDKs), as critical regulators of cell cycle progression. The CDKs are constitutively expressed throughout the cell cycle (Figure 6.2).

Figure 6.2

The eukaryotic cell cycle. M, Mitosis; $G_1 + S + G_2$, interphase; G, gap; S, DNA synthesis; G_0, terminal differentiation and senescence. In normal mammalian tissues, the cell cycle time is about 24 h. In neoplasms, the cell cycle time is very variable but is usually somewhere between 24 and 600 hours. This high variability occurs not just between different types of neoplasms but is also seen between the cells of neoplasms of the same histological type.

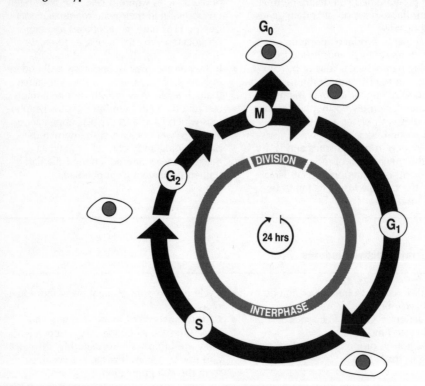

Separate cyclins and CDKs function at specific phases in the cell cycle. The G_1 phase cyclins include cyclin C, three forms of cyclin D and cyclin E. Cyclin C activates cdc2 p34 or a related CDK at the G_1 restriction point, while cyclin E is involved in the regulation of CDKs at the beginning of S phase. Of the type D cyclins, cyclin D1 corresponds to the oncogene *PRAD1*, which maps to the site of the *Bcl*1 rearrangement in some leukaemias and lymphomas (see p. 122). Different members of the cyclin D family regulate CDK4- and CDK6-mediated phosphorylation of the retinoblastoma gene product, *Rb* p110 (see p. 123), and G_1 to S progression critically involves both CDK4 and CDK6. Cyclin A associates with Cdk2 p33 and functions in S phase. G_2–M transition involves the prototype member of the CDK family, cdc2 p34–cyclin B protein kinase (MPF, maturation promoting factor), and the related protein, CDK2 (Figure 6.3).

Monoclonal and/or polyclonal antibodies have been raised against the cyclins and the CDKs and can be used for immunohistochemistry, immunoprecipitation or Western blotting (see p. 18).

6.2.2 *Proliferating cell nuclear antigen (PCNA)*

Proliferating cell nuclear antigen (also known as the polymerase δ associated protein or, confusingly, also sometimes called cyclin) is synthesised in early G_1 and reaches its maximum synthesis during the S phase of the cell cycle. In early S phase, PCNA has a very granular distribution but is absent from the nucleolus, whereas, in late S phase, PCNA is prominent in the nucleolus. PCNA is a 36 kDa protein that is highly conserved, being present in animal and plant cells and yeasts. It therefore serves as a an excellent marker of cycling cells.

6.2.3 *Ki67 (MIB1)*

Anti-Ki67 is the name given to a group of antibodies, the first of which was synthesised at the University of Kiel in Germany (hence Ki), that react with nuclear proteins (called Ki67) permanently present in cycling cells. The antibodies react with cells at all stages of the cell cycle (late G_1, S, G_2 and M phase) but not with cells in G_0 phase, and so serve as an ideal marker of the growth fraction.

The Ki67 proteins accumulate in cells during the cell cycle, and the distribution of the Ki67 proteins varies within the nucleus at different stages of the cycle. In the daughter cells following mitosis, the Ki67 proteins are present in the

Figure 6.3
**Cyclins, CDKs (cyclin-dependent kinases)
and the cell cycle. mPF, maturation-
promoting factor.**

Cyclins and the cell cycle

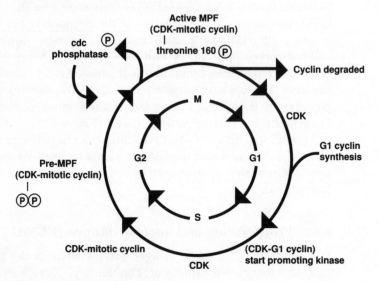

MPF - maturation promoting factor

Cyclin expression patterns

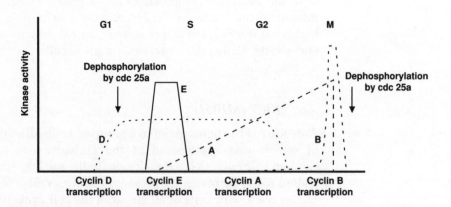

All 4 cyclins have different patterns of expression, associated with kinase activity seen.

cdc2 is present throughout the cell cycle.

perinucleolar bodies which then fuse to give the first nucleoli. These early nucleoli come together, so that their number decreases during the G_1 phase up to the G_1–S transition, giving 1–3 large, round nucleoli in S phase. During S phase, the nucleoli increase in size up to the S–G_2 transition, when the nucleoli assume an irregular outline. This irregularity is followed by the disappearance of the Ki67 proteins in the nucleoli in G_2 as the Ki67 proteins, with some other proteins, form a layer around the chromosomes during mitosis. This Ki67 protein coat covers the chromosomes, except for the centromeres and telomeres where there are no genes. The function of the Ki67 proteins is to protect the DNA of the genes from abnormal activation by cytoplasmic activators during the period of mitosis when the nuclear membrane has disappeared. If a cell leaves the cell cycle, all the Ki67 proteins disappear within about 20 min.

6.2.4 *Nucleolar organiser regions (NORs)*

Nucleolar organiser regions (NORs) are complexes of proteins and rRNA (see p. 17). They can be readily demonstrated using a silver nitrate technique which deposits black silver grains at the site of the NORs, hence their common abbreviation is AgNORs (Figure 6.4).

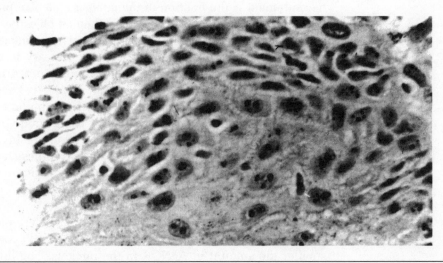

Figure 6.4
AgNOR staining of a benign squamous proliferation shows the AgNORs present as black dots within the nuclei of the keratinocytes.

AgNORs are a very useful marker of the cell cycle duration. Previously, measurement of cell cycle time needed
experiments utilizing 5-bromo-2'-deoxyuridine or tritiated
thymidine with counts being made of the number of mitoses
containing the labelled DNA precursor. When a cell needs to
synthesise proteins, it first adjusts the number of copies of
RNA and, also, the ribosomal machinery by first synthesising
the necessary ribosomal proteins. The AgNORs demonstrate
this ribogenic activity of the cell (if a typical cell contains 10^6
ribosomes and divides every 24 h, it must export 14 000 new
ribosomal subunits every minute from the nucleolus through
the nuclear pore complexes to the cytoplasm).

The number of AgNORs in a nucleus can be counted using
a light microscope. Clearly, there are differences from cell to
cell in the number and size of the AgNORs. If the AgNORs
are counted for a number of cells, say 100 cells of a particular
tumour, then a mean AgNOR score can be calculated and
compared with that in normal tissues or other tumours.
There has been extensive estimation of AgNORs in a variety
of normal and non-neoplastic tissues since the mid-1980s. In
many types of neoplasms, the AgNOR scores can be shown
to relate to other indicators of cell proliferation (e.g. Ki67
staining or tritiated thymidine incorporation) and to correlate with prognostic groupings (e.g. the division of non-
Hodgkin's lymphomas into low-grade and high-grade
groups).

There were, however, some problems with the AgNOR
technique. Some were technical, relating to the staining
methodology; others centred around the relative merits of
counting either the number or measuring the size of the
AgNORs. What has emerged as a consensus is that the best
measurement is the fraction of the nuclear area occupied by
the AgNORs. This AgNOR fraction is then an objective measurement of the ribogenic activity of a given nucleus. To
calculate the AgNOR fraction it is first necessary to determine which of the cells in a given population are in the
cell cycle (Ki67 +ve) and which are not (Ki67 −ve) because
the AgNOR fraction in cycling cells is the important measurement (this can be performed readily using computerised
image analysis). The cycling cells (Ki67 +ve) are involved
solely in preparation for mitosis and are synthesising the
enzymes necessary to replicate DNA, whilst the non-cycling
cells (Ki67 −ve) are involved in the normal function of the
particular tissue.

The intensity of the AgNOR staining is inversely proportional to the cell cycle time; so, the longer the cell cycle, the
smaller the amount of AgNOR in the nucleus. The limiting

factor in the cell cycle seems to be growth. Is the cell large enough to divide? Has it replicated enough mitochondria and other organelles and has it replicated the DNA? The faster a cell can synthesize proteins and enzymes for DNA replication, the faster it can cycle. The fraction of the nucleus that is occupied by AgNORs represents the metabolic capacity of that cell. Since this metabolic capacity represents the limiting factor for cellular replication, AgNORs represent a static measurement of cell cycle duration.

The potential for tumour growth can be described by the equation:

$$N_t = N_0 \times (P + 1)^{1/T}$$

where N_t is number of cells in a population after a certain period of time t; N_0 is number of cells in the population at the start, P is growth fraction (Ki67 index), t is period of time, T is cell cycle time, and $1/T$ is proportional to AgNOR intensity.

The AgNOR intensity is *not* predictive of prognosis in cancer patients, i.e. whether an individual patient will live or die. What the AgNOR intensity predicts is the length of time to death in those patients that do die. A high AgNOR intensity (in the Ki67 +ve tumour cells) correlates with a short time to death, whereas a low AgNOR intensity (in the Ki67 +ve tumour cells) means a long time to death. The similarities in the patients who die of a particular cancer are (1) that they have all produced the same tumour burden, and (2) that their cancer cells have all passed through the same number of cell cycles. It seems likely that some of the capabilities that cancers show as they grow and progress (e.g. metastatic potential, multidrug resistance, p53 mutations) may require a specific number of cell cycles to be attained and, furthermore, if they are not acquired within those number of cycles they will not be obtained. This is very similar to the situation in embryonic development where definitive differentiation profiles are obtained after defined numbers of cell cycles.

6.3 Gene rearrangement analysis

This technique can be applied to samples containing lymphoid cells to ascertain whether the lymphoid cells are monoclonal (and hence a neoplasm) or polyclonal (and hence a reactive proliferation) (Box 6.4). Demonstration of monoclonality within lymphoid neoplasms is a special case. It can be achieved by demonstrating, within the tumour

Box 6.4 **Clones**

- A clone is a population of cells derived from a single precursor cell.
- Clones can form part of normal or neoplastic tissues.
- Clones can be demonstrated in many normal tissues (e.g. human gastric and colonic glands).
- Clonal expansion of B and T lymphoid cells is a vital component in normal immune responses. Memory cells encountering antigens (binding via surface immunoglobulin on B cells or the T cell receptor on T cells) are stimulated to both proliferate (creating clones) and mature (producing mature effector cells, i.e. plasma cells and activated effector T cells). Large numbers of different clones are produced because natural antigens have many epitopes.

- Neoplasms are clonal proliferations. Evidence for this statement comes from a variety of sources including:

1. Neoplasms arising in females heterozygous for different isoenzymes of glucose-6-phosphate dehydrogenase (the genes are on the X chromosomes) produce only one of the isoenzymes in the cells of a given neoplasm.
2. Many neoplasms show specific chromosomal abnormalities on cytogenetic analysis, such as the translocation t(11;22)(q24;q12) in Ewing's sarcoma, the deletion del(1p36.2-3) in neuroblastoma, and the translocation t(2;13)(q35;q14) in alveolar rhabdomyosarcomas. These cytogenetic abnormalities are present in all the cells of the neoplasm.

cells, a single glucose-6-phosphate dehydrogenase isoenzyme in a suitable patient, or a specific cytogenetic abnormality such as t(8,14) or t(14,18). However, there are specific attributes of lymphoid cells that enable the use of other strategies. These are:

- The phenotypic expression of a single immunoglobulin light (Ig) chain isotype (either kappa or lambda) by the neoplasm. This is known as light chain restriction.

- The production of an anti-idiotype antibody which recognises the monoclonal immunoglobulin product of a mature B cell tumour.

- Demonstration of the same unique DNA rearrangement pattern in all cells of the clone. DNA rearrangement of genes is mandatory within cells of lymphoid lineage (B or T) and occurs early in development. In B cells the rearranged genes are the immunoglobulin genes, in T cells those that code for the T cell receptor.

Malignant tumours derived from lymphoid cells are called lymphomas and are classified as either Hodgkin's disease or non-Hodgkin's lymphoma, the latter being usually assignable to B cell or T cell lineage. Non-Hodgkin's lymphomas are common neoplasms and account for between 3% and 5% of all malignancies in the UK. Most B cell lymphomas will show a rearrangement of the genes that code for the polypeptides that will constitute the Ig light and heavy chains. Similarly, most T cell lymphomas will show a rearrangement of the genes coding for the T cell receptor. The rearrangement is unique to the cells of that particular neoplastic clone and is very different from the non-rearranged state (germ line DNA).

If we start by considering the Ig light chain genes, there are separate genes which code for the variable region (the 1000 or so V_L genes), a small joining region (the 10 J_L genes) and the constant region of the light chain (the C_κ genes on chromosome 2 and C_λ genes on chromosome 22) (Figure 6.5).

Each light chain is synthesised as a single unit by translation of mRNA carrying the code for the variable, joining and constant segments of the whole light chain. The selection of a single light chain gene therefore requires the formation and exclusion of DNA loops between the V_L and J_L genes and then the VJ and C gene segments (Figure 6.6). This occurs early in B cell development. As the V_L genes form a continuous series on the chromosome and the excluded DNA loop is of random length, so the active VJC light chain gene is of unique length. A similar rearrangement of V_H, J_H and C_H genes occurs on chromosome 14 in the selection of an active Ig heavy chain gene, with the inclusion of one of about 50 D_H (diversity) genes between the V_H and J_H genes which codes for hypervariable region III (Figures 6.7 and 6.8).

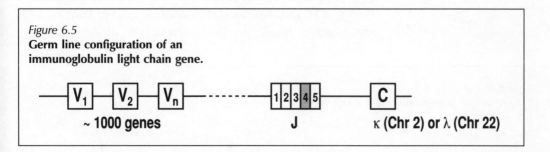

Figure 6.5
Germ line configuration of an immunoglobulin light chain gene.

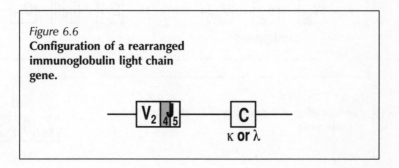

Figure 6.6
Configuration of a rearranged immunoglobulin light chain gene.

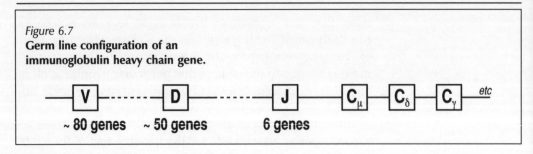

Figure 6.7
**Germ line configuration of an
immunoglobulin heavy chain gene.**

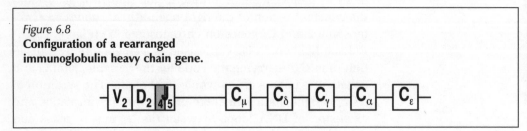

Figure 6.8
**Configuration of a rearranged
immunoglobulin heavy chain gene.**

The T cell receptor (TCR) is a two-chain surface membrane-anchored heterodimer formed of either α and β chains (the commonest form) or δ and γ chains (Figure 6.9). Each chain has variable and constant regions encoded by separate genes that rearrange in a similar fashion to the Ig genes (Figure 6.10).

To analyse Ig or TCR gene rearrangements, DNA is extracted from the lymphoid cells by standard techniques and subjected to digestion by restriction endonucleases (see p. 13). Because rearrangement of heavy, light or TCR chain genes results in the introduction of new cleavage

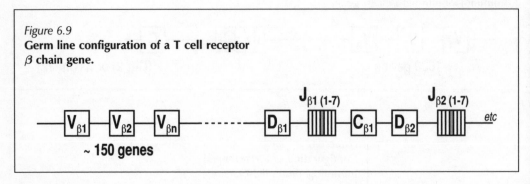

Figure 6.9
**Germ line configuration of a T cell receptor
β chain gene.**

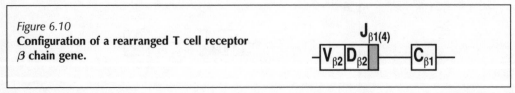

Figure 6.10
**Configuration of a rearranged T cell receptor
β chain gene.**

sites compared with the germ line configuration, the fragments produced following digestion of rearranged immuno-globulin or TCR genes will differ in length compared with the fragments produced by digestion of germ line genes. This difference in length (and hence molecular weight) can be used to separate the DNA fragments by agarose gel electrophoresis. The DNA in the gel is then denatured, which renders the normally double-stranded fragments single-stranded so that they can subsequently hybridise with complementary probes of DNA. The DNA fragments are then transferred to a nitrocellulose filter by Southern blotting (see p. 17), hybridized with labelled DNA probes (usually ^{32}P plasmid probes), and subjected to autoradiography (see p. 3). This reveals the clonal DNA as a distinct band or bands at a different position to the germ line DNA band (Figure 6.11).

There is insufficient DNA at any one location from rearranged but polyclonal lymphoid cells to form a visible band, because the radioactivity is spread all over the filter and is below the limit of resolution. Southern blotting will detect the gene rearrangements present in clonal lymphoid cells as long as they constitute more than 5% of the population in the sample. If the neoplastic clone represents a smaller proportion, then a PCR-based strategy will be needed to detect it.

Problems with interpretation of Southern blots can occur if there are DNA polymorphisms or variant bands. DNA

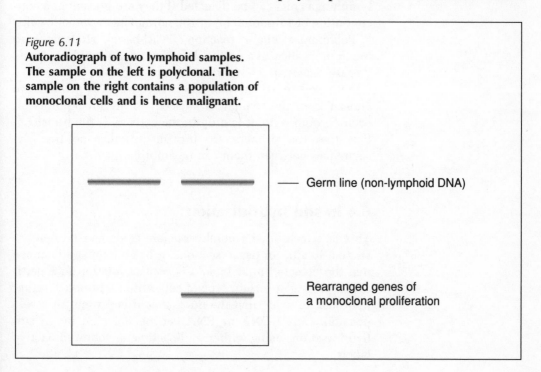

Figure 6.11
Autoradiograph of two lymphoid samples. The sample on the left is polyclonal. The sample on the right contains a population of monoclonal cells and is hence malignant.

—— Germ line (non-lymphoid DNA)

—— Rearranged genes of a monoclonal proliferation

polymorphisms occur when two or more alleles at a locus (the site of a gene on a chromosome) are relatively common in the population (see p. 156). Polymorphisms are commonest at the C_λ locus. Alleles are of different sizes and so separate to different positions (compared to the germ line) during electrophoresis. DNA polymorphisms can be distinguished from monoclonal bands because the extra bands seen on the autoradiograph of the Southern blot are shared by both the lymphoid DNA and the non-lymphoid control DNA. They can also be excluded as the cause of bands not in the germ line position by demonstrating rearrangements after digestion with at least two different restriction endonucleases in each test case. Because one or more of the three restriction enzymes typically used (*EcoRI, HindIII, BamHI*) may fail to demonstrate a rearrangement, running all three has become the universally accepted protocol. Trying to get by with less than three enzymes carries too great a risk of false-negatives or false-positives (from cross-hybridisation or partial digests).

Conventional Southern blotting can also be used for the detection of residual disease in the bone marrow after cytotoxic chemotherapy. An example would be a type of non-Hodgkin's lymphoma known as follicular lymphoma in which there is commonly a translocation between chromosomes 14 and 18. With such mutations as markers, residual lymphoma cells can be detected if they are present at a concentration of 1 tumour cell in 100 normal haemopoietic cells.

Polymerase chain reaction (PCR)-based strategies are much more sensitive for the detection of residual lymphoma disease and can be used to detect a concentration of 1 tumour cell in 10^6 normal cells. The two PCR primers are chosen from the sequences adjacent to the breakpoints on each chromosome. It is only in the cells with the translocation that the primers are brought together so that the sequences between them can be amplified.

6.4 In situ hybridisation

This is a technique whereby nucleic acids can be demonstrated 'in situ' in tissue sections. It has the great advantage that the detected nucleic acid is seen in relation to a particular cell (e.g. a certain type of cell within a tumour), rather than just knowing that the nucleic acid is present in a suspension. Either DNA or RNA can be detected by in situ hybridisation using either radioactive or non-radioactive labels.

6.4.1 *In situ hybridisation for DNA*

In situ hybridisation is the method for the detection of specific DNA sequences in tissue sections. It uses a labelled complementary nucleic acid sequence (either a cloned DNA probe or a synthetic oligonucleotide probe) as a probe to hybridise to the target DNA. Both the probe and the target DNA need to be single-stranded for hybridisation to occur.

6.4.2 *In situ hybridisation for RNA*

This is the same as the technique for DNA but is applied to the detection of RNA sequences in tissue sections.

6.4.3 *In situ hybridisation for infectious agents*

In situ hybridisation has been utilised to investigate different kinds of microorganisms in tissues, but the main use has been in the detection of viral infections. An important clinical example is the detection of human papillomavirus (HPV) subtypes in cervical cytology specimens by in situ hybridisation on routine cytology smears. HPV is a common viral infection of the metaplastic stratified squamous epithelium of the transformation zone of the uterine cervix. There are over 70 different types of HPV but only some of these (types 16, 18 and, to a lesser extent, 33) confer an increased risk of progression from cervical intra-epithelial neoplasia (cervical dysplasia) to invasive cervical squamous carcinoma. For example, DNA from HPV type 16 can be found in 47% of invasive cervical carcinomas by Southern blotting (see p. 17). HPV genomic DNA is functionally divided into early (E) and late (L) genes. E genes encode for DNA replication, transcriptional regulation and transformation. These early genes evolved to remove blocks on DNA synthesis in quiescent cells so allowing viral DNA replication. The L genes encode the major and minor capsid proteins. E gene products act as oncoproteins, inactivating the tumour suppressor gene products p53 and p110 (see p. 123) and others, which leads to uncontrolled cell proliferation. Identification of women infected with the particular oncogenic HPV types is clearly important as it could then allow closer clinical follow-up.

In situ hybridisation techniques are also useful for the identification of Epstein–Barr virus (EBV) and CMV infections in tissue samples, especially in immunosuppressed patients. In situ hybridisation has the advantage over immunohistochemistry that it can detect latent infections as well as productive infections. In transplant patients (liver, bone marrow, etc.) and in patients with AIDS, EBV-driven lymphoproliferations can be a serious, sometimes fatal, occurrence. Some of these are polyclonal and some are monoclonal, but the clonality actually makes little difference to the seriousness of the disease. Monoclonal EBV-driven lymphoproliferations can appear histologically identical to non-Hodgkin's lymphomas. Treatment for EBV-driven lymphoproliferations in transplant patients is usually by reduction of the immunosuppression; this may lead to immune-mediated damage, and even rejection, of the graft, so it is a very important and difficult decision to make.

In situ hybridisation has an advantage over PCR in some of these situations. PCR is so sensitive it can pick up subclinical infections which may not matter clinically.

6.5 Oncogenes and tumour suppressor genes

Oncogenes are genes whose changed expression or altered product is essential to the production or maintenance of cancers (Box 6.5). Novel oncogenes may be created by the cytogenetic rearrangements present in some tumours. Examples are the *Bcl*1 and *Bcl*2 rearrangments found in some leukaemias and lymphomas, and the *bcr-abl* oncogene created when c-*abl* is translocated from chromosome 9 to chromosome 22 in Philadelphia chromosome positive chronic granulocytic leukaemia.

Although an understanding of which oncogenes may be activated in certain cancers is critical in understanding cancer biology, there is no really good evidence that knowing the oncogene status provides any additional prognostic information in common cancers, such as breast cancer,

---

Box 6.5 **Oncogene terminology**

- v-oncogene (viral oncogene). An oncogene carried by or derived from a virus. Both DNA and RNA viruses (retroviruses) can carry oncogenes. In DNA viruses they are true viral genes and the gene product is needed by the virus. In RNA viruses they have been acquired when an antecedent of the virus, within an infected human or primate cell, incorporated a c-oncogene into the viral genome. v-oncogenes in retroviruses have no viral product. Infection of a human cell by either type of virus can lead to the introduction of the viral oncogene.

- c-oncogene (cellular oncogene). A gene within the normal human genome that is normally not expressed in adults. The gene product may have been required during embryogenesis or fetal life but not subsequently. Expression of the gene, with altered or permanent growth signals, may lead to cancer.
- Proto-oncogene. The non-activated state of a c-oncogene.
- Anti-oncogene. A tumour suppressor gene (see p. 123).

over and above the usual prognostic indicators of tumour size, lymph node status and mitotic activity index. Unfortunately, this is probably true for all human neoplasms at present but the situation may change.

6.5.1 Tumour suppressor genes

The protein products of tumour suppressor genes (also known as anti-oncogenes) are involved in apoptosis (see p. 109) and in the negative control of cell cycle proliferation (whereas oncogenes have a dominant action and exhibit positive control). Damage to tumour suppressor genes leads to abnormal proteins. This can contribute to tumorigenesis because control of some aspect of proliferation or cell death is then lost.

The two most widely studied tumour suppressor genes are *p53* and the retinoblastoma susceptibility gene, *Rb*. The *p53* gene encodes a tumour suppressor protein, also called p53, which exists as either a homodimer or homotetramer. The *p53* gene can be functionally inactivated by mutation (this is a dominant negative mutation where the mutant protein complexes with and inactivates the wild type protein), by DNA tumour virus-encoded proteins (e.g. the SV40 large T antigen or adenovirus E1B and papilloma virus E6 proteins) sequestering wild type *p53*, or by interacting with an oncogene encoded protein such as the *MDM2*-oncogene encoded p90 protein. *p53* mutations have been documented in many human neoplasms (e.g. breast cancer) and there is an increased concentration of mutant p53 protein in the tumour cells.

The retinoblastoma susceptibility gene, *Rb*, encodes a phosphorylated protein, p110, that can also be inactivated by mutations or by binding to DNA tumour virus-encoded proteins (e.g. adenovirus E1A). p110 functions by attaching to and regulating the transcription factors that bind to DNA (Box 6.6).

Box 6.6 **Transcription factors**

- Transcription factors are proteins that bind selectively to specific or consensus sequences of DNA. They regulate the transcription of RNA from the corresponding DNA sequence by either promoting or preventing expression.

- The transcription factors themselves are regulated either by phosphorylation (by protein kinases) or by chemical oxidation of specific amino acid residues within their DNA binding domains.

6.6 PCR on tissue sections

It is possible to extend the PCR (polymerase chain reaction) analysis (see p. 11) to non-disrupted cells and tissues at the light-microscopic level to localise nucleic acids in tissue sections. This can be done with either radioactive or non-radioactive labels.

6.6.1 *PCR in situ hybridisation*

PCR in situ hybridisation is the PCR amplification of DNA or mRNA in tissue sections followed by in situ hybridisation (using a radioactive or non-radioactive labelled probe) to detect the amplified product (Figure 6.12).

6.6.2 *In situ PCR*

In situ PCR is the technique of PCR applied to tissue sections to localise DNA or mRNA, using either a labelled primer or oligonucleotide in the reaction mix (Figure 6.12). It is particularly useful for looking at low-frequency message expression in mixed populations of cells, such as viral targets or growth factors in tumour cell lines. Some biosystems companies now produce dedicated systems for performing in situ PCR.

6.6.3 *In situ reverse transcriptase (RT) PCR*

In situ RT-PCR is the technique for in situ PCR applied to the localisation of mRNA. It is a highly sensitive technique that can be applied to tissue sections (usually formalin-fixed paraffin sections; cryostat sections are not very successful) and allows the correlation of molecular and histological findings (Figures 6.13 and 6.14). Deparaffinised tissue sections are digested with a protease (the digestion time is optimised according to the length of time of fixation in formalin) followed by digestion with RNase-free DNase to eliminate any non-specific signal. Using specific primers and reverse transcriptase (RT), specific cDNA molecules are produced. In situ PCR is then carried out by direct incorporation of labelled (e.g. digoxigenin) deoxynucleotide triphosphates using specific primers. To ensure there is no DNA-derived signal, a second section on the same slide is treated with DNase but not RT; and to ensure that the PCR and detection steps are optimal, a third section on the same slide is not given the DNase digestion step.

Figure 6.12
Schematic representation of PCR in situ hybridisation (PCR-ISH) (above) and in situ PCR (below). For PCR-ISH, the PCR reaction mix is placed onto the tissue section and covered with a coverslip. Following amplification, a labelled probe is added and the product is developed as for conventional ISH. For in situ PCR, the PCR reaction mix contains labelled nucleotide or primer, which is placed onto the tissue section and covered with a coverslip. Once again, following amplification, the products are developed using standard immunocytochemical techniques. From: J.J. O'Leary, R. Chetty, A.K. Graham and J. O'D. McGee, 1996. *In situ PCR: Pathologist's Dream or Nightmare?* Copyright © 1996 by John Wiley & Sons Ltd. Reprinted by permission of John Wiley & Sons Ltd.

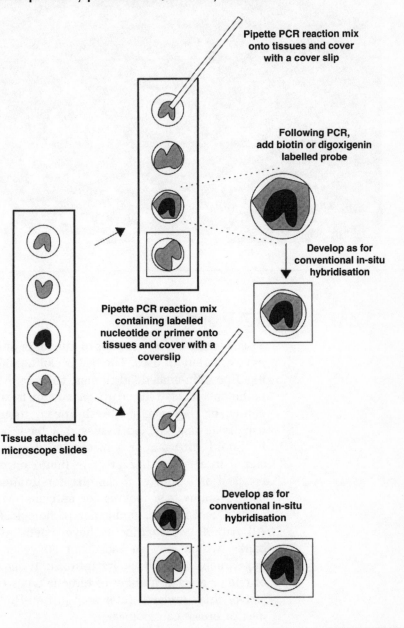

Figure 6.13
Vitamin D receptor mRNA demonstrated in human liver by in situ RT-PCR. Following amplification of the vitamin D receptor mRNA by RT-PCR, the product is detected using a ^{35}S-labelled probe. Courtesy of Jane Codd, King's College Hospital, and Andrew Mee, Manchester Royal Infirmary.

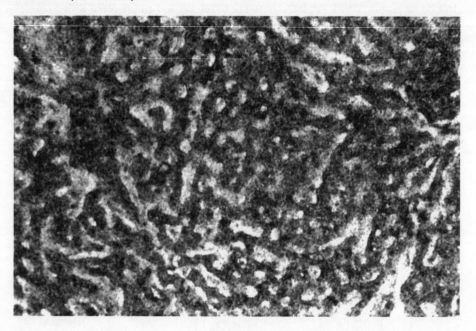

6.7 Tissue banks

Tissue banks are repositories of human or animal tissues that have a variety of uses. The banks can be either of formalin-fixed paraffin-embedded tissues, such as the surgical histopathology block files of diagnostic histopathology departments, or they can be 'fresh-frozen' tissue banks which usually means unfixed tissues that have been snap-frozen in liquid nitrogen or supercooled isopentane and then stored in a $-70°C$ freezer or in liquid nitrogen. Either can be used as a source of samples for immunohistochemical investigations (e.g. looking for antigens or other proteins), or as a resource for molecular pathological investigations. All pathology departments have extensive paraffin tissue banks, which may go back over 50 years or more. Large departments may have fresh-frozen tissue banks, usually of one or a small number of tumour types that are the subject of their research interest (e.g. non-Hodgkin's lymphomas or breast carcinomas).

Figure 6.14
**Adrenomedullin mRNA demonstrated in neurons of rat brain by in situ RT-PCR.
Digoxigenin-labelled deoxynucleotide triphosphates are directly incorporated into the PCR
product and are then detected using an anti-digoxigenin antibody conjugated to alkaline
phosphatase. Courtesy of Jane Codd, King's College Hospital.**

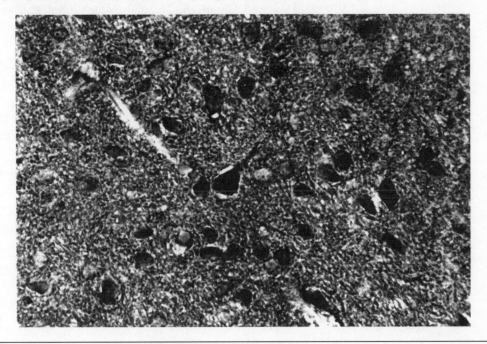

Recently, tissue banks have become the subject of some
ethical concern. What rights would an individual patient
have, for example, if genetic material extracted from a specimen of his tissues was the foundation of a test that had commercial application? Or what if scientists found a gene in
that tissue that could lead to disease? Should the patient be
traced and told? Even if there were no treatment? These
questions are now being addressed, ethically and construc-
tively, particularly in the USA. There may have to be
changes in the consent form that patients sign when they
are about to have an operation. In the UK it has always
been implied, and expected by the patient, that tissues
removed by the surgeon will, if necessary, be examined by
a pathologist to determine the nature of the disease(s) present
and the completeness of its excision. Postmortem consent
forms usually have a section where specific consent, that
tissues can be used for teaching or research, can be given
or withheld. Operative consent forms may need to have a
similar section in the near future.

Further reading

Alberts, B., Bray, D., Lewis, J., Raff, M., Roberts, K. and Watson, J.D., 1989. *Molecular Biology of the Cell*, 2nd edn. New York: Garland Publishing.

Brown, D.C. and Gatter, K.C., 1990. Monoclonal antibody Ki67: its use in histopathology. *Histopathology*, **17**, 489–503.

Franks, L.M. and Teich, N.M., 1991. *Introduction to the Cellular and Molecular Biology of Cancer*. Oxford: Oxford University Press.

Herrington, C.S. and McGee, J.O'D., 1992. *Diagnostic Molecular Pathology: A Practical Approach*. Oxford: IRL Press.

Nuovo, G.J. (Ed.), 1992. *PCR In Situ Hybridisation: Protocols and Applications*. New York: Raven Press.

O'Leary, J.J., Chetty, R., Graham, A.K. and McGee, J.O'D., 1996. In situ PCR: pathologist's dream or nightmare? *Journal of Pathology*, **178**, 11–20.

Stevens, A. and Lowe, J. 1992. *Histology*. London: Gower Medical.

Van Diest, P.J. and Baak, J.P.A., 1992. *Quantitative Cyto- and Histoprognosis in Breast Cancer*. Amsterdam: Elsevier, p. 182.

7 The Metabolic Basis of Inherited Disease

William J. Marshall

7.1 Introduction

Some 5000 diseases have been described which have their origins in one of the 50 000–100 000 genes estimated to comprise the human genome. Individually, these diseases are uncommon; indeed, many are very rare, but, overall, approximately 1% of babies are born with an inherited metabolic defect attributable to a mutation in a single gene and collectively these conditions are an important cause of morbidity and mortality, particularly in children, in whom they often—but by no means always—first come to notice.

This chapter opens with a brief reminder of some aspects of molecular genetics relevant to inherited molecular diseases. There follow discussions of the relationships between the basic defects and their clinical expression; the techniques used for diagnosis and for detection before clinical presentation (screening); and of the approaches to treatment, illustrated throughout by examples relating to specific conditions. No attempt is made to provide a comprehensive description of any one, or group of, conditions. Readers requiring further information may consult the reference texts given at the end of the chapter.

The concept of inherited metabolic disease dates from the early years of the twentieth century, and the studies by the English physician Sir Archibald Garrod on alkaptonuria, a familial condition characterised by the passage of urine that turns dark on standing, pigmentation in cartilage and soft tissues, and degeneration of articular cartilage leading to arthritis.

The urine in patients with alkaptonuria contains excessive quantities of homogentisic acid, a metabolite of tyrosine, and Garrod proposed that this was due to decreased activity or absence of an enzyme responsible for the metabolism of homogentisic acid. This hypothesis was only proved correct in 1958 when La Du reported a deficiency of the enzyme homogentisic acid oxidase in patients with alkaptonuria.

The concept of 'one gene, one enzyme' (implying that individual enzymes are coded for by single genes) was

developed in the 1940s and 1950s by Beadle and Tatum. The
first inherited disease proven to involve an abnormal protein
was sickle cell disease, an inherited haemolytic anaemia. In
1949, Pauling and his colleagues demonstrated that this is
due to an abnormal haemoglobin, and, in 1956, Ingram
reported that the molecular basis of this is a single amino
acid substitution in the haemoglobin molecule (valine for
glutamine at position 6 in the β chain).

A considerable number of inherited disorders of haemoglobin, both qualitative and quantitative, are now known.
They are discussed separately in Chapter 4 of this book
and will not be discussed further here.

Beadle and Tatum's simple concept has required considerable revision over the years, particularly since the advent
of molecular genetic techniques for studying the molecular
basis of inherited diseases. It is now known that genes can
code for proteins other than enzymes (e.g. structural proteins
such as collagen) and that some enzymes are the products of
more than one gene. Alteration of the activity of a gene can
give rise to different gene products, while the polypeptide
products of some genes can be modified post-translationally
to give more than one product. In each of these instances,
examples of genetically determined abnormalities of the process have been described which lead to disease.

It has also become clear that the 'central dogma' of molecular biology as espoused in the 1960s, that DNA codes for
RNA which codes for protein and that the reverse cannot
happen, is untrue. Indeed, enzymes known as reverse tran-
scriptases, capable of synthesising DNA from RNA, are an
important tool of the molecular biologist (see p. 6). But it
remains true that, in eukaryotic organisms like man, the
genetic information resides in DNA, and that mutations in
DNA are the basis of inherited diseases, whether obviously
metabolic in the sense that they involve an enzyme in a
metabolic pathway and give rise to a distinct biochemical
syndrome, or involve structural, transport or other proteins.

7.2 The nature and pathogenesis of inherited metabolic disease

7.2.1 Diseases due to defective enzymes

Consider a hypothetical metabolic pathway leading from a
precursor substrate 'A' (see Figure 7.1) through various intermediates, 'B', 'C' and 'D' to a final product, 'E', each step of
which is catalysed by a specific enzyme, 'a', 'b', etc. In the

pathway in Figure 7.1, a second, but quantitatively less important pathway, involves the transformation of 'C' to product 'D'' by the enzyme 'c''.

Consider now an inherited defect in which the enzyme 'c' is absent (or, as more frequently happens, is present but has reduced activity). This will decrease the formation of the product 'D'. The clinical consequences of an inherited metabolic defect are frequently related to decreased formation of an essential product; for example, glucose 6-phosphatase deficiency blocks the formation of glucose from glycogen (Figure 7.2) and causes hypoglycaemia in the fasting state.

Glucose 6-phosphatase deficiency also illustrates another potentially harmful consequence of decreased enzyme activity, namely, accumulation of a substrate 'upstream' of the enzyme. Glucose 6-phosphatase deficiency is one of the glycogen storage diseases, all of which are inherited disorders affecting enzymes required for normal glycogen metabolism, defects in which lead to excessive accumulation of glycogen. When this occurs in the liver, it can give rise to hepatic enlargement and liver failure. Lactic acidosis (due to increased glycolysis) is another predictable consequence of glucose 6-phosphatase deficiency.

Phenylketonuria is another example of an inherited metabolic disease in which accumulation of a substrate, this time of the defective enzyme itself, occurs with harmful consequences. Classical phenylketonuria is due to decreased activity of the enzyme phenylalanine hydroxylase, which catalyses the formation of tyrosine from phenylalanine (Figure 7.3). Both tyrosine and phenylalanine are amino acids present in many foodstuffs, so the decreased formation of tyrosine may not give rise to harm, but phenylalanine is neurotoxic in high concentrations and its continued inges-

---

Figure 7.1
A hypothetical metabolic pathway in which substrate A is converted by sequential enzyme-catalysed steps through intermediates B and C to the major product D, or a minor product, D'. Enzymes are indicated by lowercase letters.

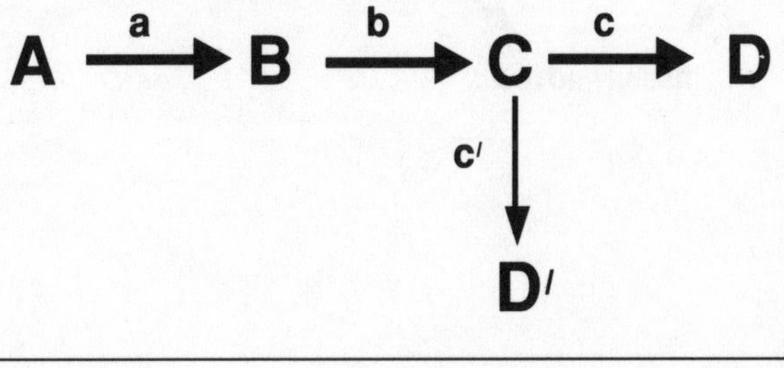

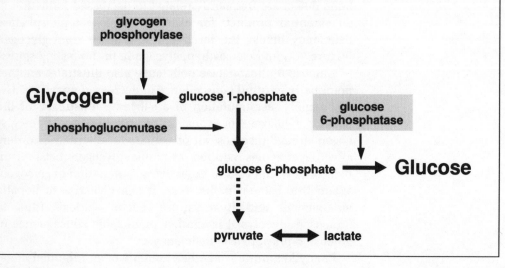

Figure 7.2

Glycogen metabolism. Glycogen is phosphorylated to glucose 1-phosphate by glycogen phosphorylase, and this intermediate is converted to glucose 6-phosphate by phosphoglucomutase. Glucose 6-phosphate can be metabolised through the glycolytic pathway to pyruvate and lactate, or converted to glucose by glucose-6-phosphate. Deficiency of this enzyme leads to defective glucose synthesis and hypoglycaemia in the fasting state, while increased glycolysis causes lactic acidosis.

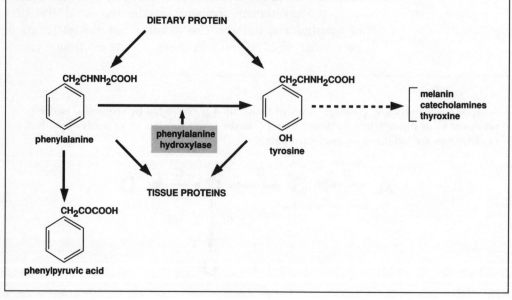

Figure 7.3

Phenylalanine metabolism. The enzyme phenylalanine hydroxylase converts phenylalanine to tyrosine. Deficiency of the enzyme leads to accumulation of phenylalanine and increased formation of the normally minor product, phenylpyruvic acid.

tion leads to high plasma concentrations and severe damage to the brains of affected babies. Fortunately, this can be averted by early detection and appropriate treatment, as is discussed later in this chapter (p. 148).

The name, phenylketonuria, illustrates a further potential consequence of an inherited enzyme defect—the increased formation of a normally minor metabolic product. Looking back at Figure 7.1, it can be inferred that, if enzyme 'c' is defective, there may be accumulation of metabolite 'C', leading by a mass action effect to increased formation of product 'D'. Returning to the example of phenylketonuria, under normal circumstances, the major route for the metabolism of phenylalanine (unless it is incorporated into protein) is conversion to tyrosine. If this pathway is blocked, there is increased formation of phenylpyruvic acid, normally only a minor product, and this is excreted in the urine (where its detection used to provide the basis of a screening test for the disease).

In addition to classic phenylketonuria, two other inherited metabolic disorders cause hyperphenylalaninaemia. These are dihydropteridine reductase (DHPR) deficiency and tetra-hydrobiopterin (BH₄) synthase deficiency. BH₄ is a cofactor for phenylalanine hydroxylase (see Figure 7.4), and both

Figure 7.4
The role of tetrahydrobiopterin (BH₄) in phenylalanine metabolism. BH₄ is a cofactor for phenylalanine hydroxylase, acting as an electron carrier, and is converted in the reaction to quinonoid dihydrobiopterin (qBH₂). BH₄ can be synthesised de novo from guanidine triphosphate (GTP) and regenerated from qBH₂ in an NADH-linked reaction catalysed by dihydropteridine reductase (DHPR). Either defective synthesis or regeneration of BH₄ can reduce phenylalanine hydroxylation and cause hyperphenylalaninaemia.

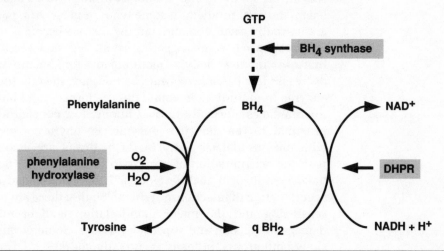

these defects interfere with the conversion of phenylalanine to tyrosine by reducing the availability of the cofactor. BH_4 is also a cofactor for the synthesis of the neurotransmitters dopamine and serotonin, with the result that the clinical presentations of DHPR and BH_4 synthase deficiency differ from that of classical phenylketonuria despite the fact that hyperphenylalaninaemia is common to all three. Infants with DHPR and BH_4 synthase deficiency present with hypotonia and epileptic fits. The neonatal screening test detects the cofactor defects as well as classic phenylketonuria, although the tests required for confirmation are different.

Metabolic pathways are frequently subject to control by negative feedback. Typically, the enzyme catalysing one step is 'rate-limiting'—i.e. its activity is less than that of the other enzymes and so controls the rate of the pathway over-all. Negative feedback involves the product of the pathway inhibiting the rate-limiting enzyme and so controlling the rate of its own formation. If, due to a deficiency of an enzyme acting 'downstream' of the rate-limiting step, there is decreased formation of product, the inhibition of the rate-limiting enzyme may be released. This may greatly increase the formation of potentially toxic precursors and increase the formation of normally minor metabolites.

This is illustrated by a group of inherited metabolic disorders called the porphyrias. The synthesis of haem, the iron-containing, oxygen-binding tetrapyrrole constituent of haemoglobin, requires a sequence of enzyme-catalysed reactions leading from succinyl-CoA and acetyl-CoA to the end-product (Figure 7.5). Diseases are known due to deficiency of each of the enzymes in the pathway. The rate-limiting enzyme, δ-aminolaevulinic acid synthase, catalyses the first step in the pathway and is subject to negative feedback by haem, the final product. In some types of porphyria, patients are normally asymptomatic, but the disease becomes apparent in reponse to some triggering event. One such event is an increase in the body's requirement for haem, which decreases free haem levels and, by releasing the rate-limiting enzyme from inhibition, stimulates the formation of harmful intermediates, normally present in only low concentrations, proximal to the defective enzyme. Porphyrias associated with enzyme defects early in the pathway are associated with the accumulation of porphyrin precursors and tend to have neurological manifestations. Later defects cause the accumulation of intact porphyrin molecules; these are photosensitising and the major manifestation is light-induced damage to the skin and superficial tissues. Some porphyrias show both groups of features. Broadly speaking, the photo-

Figure 7.5

The biosynthesis of porphyrins. PBG deaminase is also known as hydroxymethylbilane synthase, and ALA dehydratase as PBG synthase. The porphyrias caused by deficiency of each of the enzymes are shown in parentheses underneath the name of the enzyme.

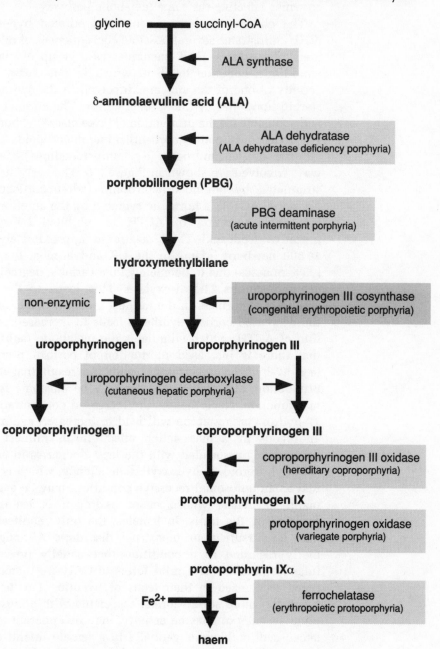

sensitising porphyrias tend to be more chronic, whereas the neurological porphyrias may become clinically apparent only when some precipitating event occurs. The porphyrias thus illustrate that there can be both similarities and differences in the manifestations of inherited diseases involving enzymes affecting the same metabolic pathway.

The clinical syndrome, congenital adrenal hyperplasia (CAH), illustrates several possible consequences of inherited metabolic defects. CAH encompasses a group of inherited conditions, common to all of which is a decrease in the activity of one of the enzymes involved in the synthesis of steroid hormones by the adrenal cortex. The adrenal cortex is responsible for the production of three classes of hormone: glucocorticoids (mainly cortisol), mineralocorticoids (mainly aldosterone), and androgens (e.g. androstenedione). The pathways involved are shown in Figure 7.6. The early steps are stimulated by adrenocorticotrophin (adrenocorticotrophic hormone, ACTH), a hormone secreted by the anterior pituitary. The secretion of ACTH is inhibited by cortisol (negative feedback). CAH occurs in approximately 1 in 10 000 newborn infants in the UK and is most frequently (95% of cases) due to deficiency (to a variable degree) of the enzyme steroid 21-hydroxylase. This leads to decreased synthesis of cortisol. In the mildest forms of this condition, the decreased cortisol synthesis leads to increased production of ACTH, which stimulates the metabolic pathway so that there is increased production of cortisol precursors (e.g. 17-hydroxyprogesterone) with the result that cortisol production, although decreased, may be adequate for normal function. Nevertheless, the increased concentrations of 17-hydroxyprogesterone will lead to increased synthesis of androgens by a mass-action effect. In its mildest forms (presumably associated with the least decreases in enzyme activity), steroid 21-hydroxylase deficiency, which is inherited as an autosomal recessive condition, may be asymptomatic in males, whose major androgen is testosterone secreted by the testis. In females, the only manifestations may be hirsutism or menstrual disorders. Although formerly regarded as a condition that usually presents in infancy, patients with mild forms of CAH are increasingly being recognised in their teens or twenties. On the other hand, the more severe forms, associated with almost complete absence of enzyme activity, may be apparent at birth because of ambiguous genitalia in a female infant, or collapse with hypotension due a salt-losing crisis in either sex as a result of a lack of aldosterone. Steroid 21-hydroxylase is an essential enzyme in aldosterone synthesis, but only

Figure 7.6
The biosynthesis of adrenal steroid hormones. The secretion of cortisol is controlled through negative feedback by pituitary adrenocorticotrophin (ACTH). Cortisol inhibits ACTH secretion, which normally stimulates the conversion of cholesterol to pregnenolone and maintains the supply of substrate for cortisol synthesis. Deficiency of 21-hydroxylase leads to decreased synthesis of cortisol, accumulation of 17-hydroxyprogesterone and increased androgen synthesis. In some cases of deficiency of this enzyme, aldosterone synthesis is also decreased. 3HDI is the enzyme 3β-hydroxydehydrogenase, Δ^5 isomerase.

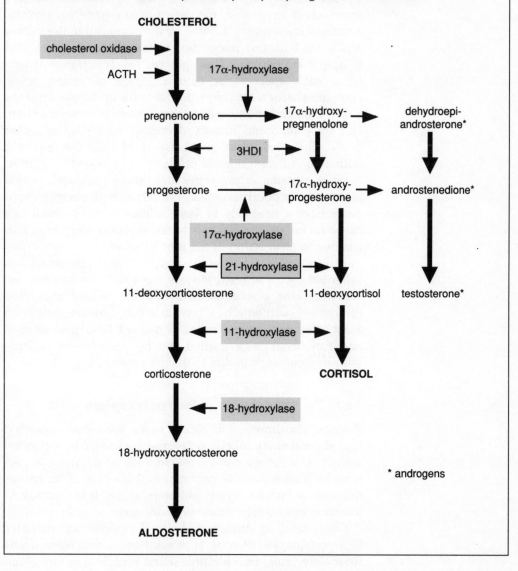

about one-third of patients with CAH due to a defect in this enzyme have a significant decrease in aldosterone synthesis. CAH thus illustrates a point made earlier — that the clinical manifestations of inherited metabolic diseases, even when due to an abnormality of the same enzyme, can be very variable, in part because of the multiplicity of mutations than can affect the gene and thus the amino acid sequence, and in turn the structure and function of the enzyme.

7.2.2 *Diseases due to defective transport proteins*

By no means all inherited metabolic diseases involve enzymes. In cystinuria, the defect is in a transport protein. Amino acids are freely filtered by the glomeruli in the kidney and are extensively reabsorbed in the proximal convoluted tubule of the nephron. This process involves transport proteins, each specific for certain groups of amino acids. Cysteine shares a transport protein with ornithine, arginine and lysine, and in cystinuria there is defective reabsorption, and thus abnormal urinary excretion, of all these amino acids. Their loss from the body is of little consequence with a normal diet; it is the presence of increased concentrations of cysteine in the urine that causes problems, as this amino acid is poorly soluble. As a result, patients with cystinuria have a tendency to form urinary calculi, which can cause severe pain (renal colic) and, exceptionally, renal failure due to obstruction of the flow of urine.

In contrast to cystinuria, isolated renal glycosuria, an inherited defect of renal glucose transport, is a benign condition because glucose is highly soluble. Indeed, it is often discovered incidentally. Glycosuria can, however, occur in association with other, more significant tubular disorders, and can occur with normal tubular function in patients with hyperglycaemia due to diabetes mellitus.

7.2.3 *Diseases due to defective receptors*

Proteins also function as receptors for hormones and other ligands, and many inherited diseases are known in which the activity of a receptor is decreased. One of the most important, both because it is common and because of its consequences, is familial hypercholesterolaemia. It is also one of the most extensively studied conditions.

Cholesterol is transported in the blood in particles (lipoproteins, see Chapter 5) in association with other lipids (triacylglycerols, phospholipids) and proteins (apoproteins). The major cholesterol-carrying lipoprotein is low-density lipoprotein (LDL), each particle of which contains a single molecule of apoprotein B-100. Cellular uptake of cholesterol involves binding of LDL to a receptor which recognises apoprotein B-100. Decreased uptake due to a decrease in the number or activity of LDL receptors on cells leads to an increase in plasma LDL. This is of significance because high levels of LDL cholesterol are causally related to increased risk of coronary heart disease. This condition is described in more detail in Chapter 5.

7.2.4 *Diseases due to defective structural proteins*

The most abundant proteins in the body are the collagens, which are structural proteins consisting of triple helices of polypeptides. Sixteen collagens are known, encoded by 28 genes, and several inherited genetic diseases of collagen have been described. Osteogenesis imperfecta, a disorder characterised by extreme fragility of bone, is due to decreased synthesis of normal type I procollagen, a precursor of the mature collagen of osteoid, the proteinaceous matrix on which hydroxyapatite is laid down. More than 100 mutations in the collagen genes *COL1A1* and *COL1A2* have been described and give rise to four main distinct phenotypes of varying severity.

The Ehlers–Danlos syndromes are a second large group of inherited disorders of collagen synthesis, characterised by skin fragility and hyperextensibility and hypermobility of joints. Affected children are very susceptible to trauma. There is no specific treatment for any of the inherited diseases of collagen synthesis.

7.3 Diagnosis of inherited metabolic disease

There are no short-cuts in the diagnosis of inherited metabolic diseases. Many of these conditions are rare, and relatively few clinicians see sufficient patients to acquire a familiarity with more than a few. Even with the commoner conditions, such as the haemoglobinopathies, cystic fibrosis and some of the dyslipidaemias, most patients are cared for by specialists. For the less common conditions, diagnosis requires a high index of suspicion, considerable clinical acumen and access to comprehensive laboratory services.

Nevertheless, most inherited metabolic diseases affecting enzymes (and these constitute the majority) can be classified into a small number of groups depending on their clinical presentation. One such classification has been attempted by Saudubray and Charpentier. They make a major division between conditions affecting one functional or anatomical

system and having a fairly uniform presentation (e.g. cystinuria) and those affecting either one metabolic pathway common to many cells or organs, or only one organ but with
widespread consequences. The latter two can be further
divided into three groups:

1. Conditions involving the synthesis of complex mole-
    cules, with irreversible clinical consequences, whose
    manifestations are not affected by food intake; examples
    include some of the glycogen-storage diseases, disorders
    characterised by the accumulation of complex lipids,
    and many others.
2. Conditions in which patients become acutely or chroni-
    cally intoxicated by the accumulation of metabolic inter-
    mediates but are symptom-free unless such intoxication
    occurs, many of which respond to dietary manipulation;
    examples include disorders of urea synthesis (causing
    hyperammoniaemia) and some disorders of amino acid
    metabolism.
3. Conditions associated with disordered energy metabo-
    lism; for example, disorders of carbohydrate metabolism
    leading to hypoglycaemia or lactic acidosis and failure to
    thrive, and disorders of electron transport in muscle
    causing myotonia.

Thus the clinician needs to be alert to the possibility of
inherited metabolic disease in patients with a wide variety of
clinical presentations, although, once the possibility has
been recognised, only a relatively small number of tests
may be required to provide reasonable confirmation of the
suspicion and point the way to a definitive diagnostic test.
This will often involve attempting to measure the affected
enzyme in tissue obtained by biopsy.

A family history of an inherited metabolic disease or of
unexplained death in childhood, or the presence of con-
sanguinity in the parents, are also strong clues to a possible inherited metabolic disease. Diagnosis is important, not
only for the management of the affected individual, but for
genetic counselling. Even where an infant dies of a suspected metabolic disease without a precise diagnosis having been made in life, it is important that every effort
should be made to make a firm diagnosis after death, to
which end, blood, urine and any tissues samples from the
child must be preserved. Skin should always be collected
for possible fibroblast culture and analysis of enzyme
activity.

7.4 Screening for inherited metabolic disease

Screening implies attempting to identify disease before it is clinically apparent. The rationale for screening is that early diagnosis should have potential for improving the outcome, for example by providing a basis for treatment which ameliorates the consequences of the conditon. For particularly severe diseases, making the diagnosis antenatally may allow the parents to choose to have the pregnancy terminated. In addition, screening for carriers (heterozygotes for recessive conditions) may provide valuable information for genetic counselling of prospective parents.

This is an area that raises major ethical and practical issues. Some inherited diseases present late: Huntington's disease, a condition which usually only becomes apparent in the fourth decade of life, causing a progressive movement disorder, dementia and eventually death, is the best known example. This is an autosomal dominant condition. Identification of the defective gene might help affected individuals to decide whether or not to have children, but also provides them with the knowledge that they will develop a rebarbative, incurable and ultimately fatal illness. The situation is further complicated by the fact that, although Huntington's disease shows complete penetrance, expressivity is age dependent. Some individuals do not develop symptoms until the seventh decade of life.

Many common conditions, including for example several cancers and coronary heart disease, are known to have a familial incidence even though they are not inherited as single gene defects. Technological advances are making it increasingly possible to predict genetic susceptibility to such diseases. Such information may be of value to an individual in helping them to avoid, for example, known environmental triggers for the disease, although it could also impose a considerable psychological burden. Furthermore, should such information be available to others (e.g. potential employers or life insurance companies), the consequences for the individual could be extremely disadvantageous.

7.4.1 Neonatal screening

In developed countries, there is virtually universal screening for phenylketonuria. Untreated, this condition, which occurs in the UK in approximately 1 per 10 000 live births, causes irreversible impairment of central nervous system development such that affected individuals are rarely able to live independently and require long-term care. Treatment (see p. 148) is by provision of a diet low in phenylalanine; this reduces the build-up of phenylalanine in the blood. Adequately treated children grow up with near normal intellectual attainment.

Screening involves measuring blood phenylalanine concentration in a sample of capillary blood collected, usually by the midwife, 6–10 days after birth. If the concentration is elevated, a further sample is obtained for confirmation. A firm diagnosis can usually be made by demonstrating unequivocal hyperphenylalaninaemia and a normal or reduced plasma tyrosine concentration. Assay of enzyme activity (which would require liver biopsy) is not necessary provided that BH_4 (tetrahydrobiopterin) deficiency (see p. 133) can be excluded.

It is technically possible to screen for many other inherited metabolic diseases at birth, but the potential benefits of any screening programme should be carefully assessed. The generally accepted criteria against which a potential screening programme should be evaluated are indicated in Box 7.1. Unfortunately, many conditions are too rare for screening to be economically viable, while, for many others, there is no effective treatment even if a suitable test is available. The only other condition for which screening with a laboratory test is well established is congenital hypothyroidism. This is usually due to a developmental abnormality, not a genetic defect, but it satisfies the criteria for neontal screening: congenital hypothyroidism is common (approximately 1 in 2500 births); late diagnosis leads to irreversible harm to the infant; early treatment is clearly beneficial, and a simple, safe, cheap and reliable screening test (measurement of thyrotrophin concentration) in a capillary blood sample is available.

Pilot programmes have been established for a variety of other disorders including galactosaemia, congenital adrenal hyperplasia and familial hypercholesterolaemia.

Box 7.1 **Criteria that must be satisfied for neonatal screening for inherited metabolic disease**

- Condition is fatal or leads to severe disability if untreated
- Treatment is available which improves outcome
- Condition is relatively common
- Availability of an acceptable, reliable, cheap, screening test (no false-negatives; some false-positives may be acceptable)

7.4.2 *Antenatal screening*

The rationale of screening for disease before birth is that detection of an affected fetus would allow the parents to consider having the pregnancy terminated. Thus the disease must be sufficiently severe for termination to be an acceptable response to its diagnosis and the diagnosis must be made reliably and sufficiently early for termination to be feasible.

Various techniques have been used for antenatal diagnosis, including: the measurement of metabolites (derived from the fetal circulation) in maternal plasma; direct measurements on fetal blood obtained by cordocentesis (sampling from the umbilical cord under ultrasound control), and assay of metabolites and enzymes in cultured amniotic fluid cells (which are of fetal origin). The first of these can only give an indication of the possible presence of an affected fetus, and is applicable to relatively few conditions (particularly some organic acidaemias), but the other techniques can give more definitive information. However, the techniques for obtaining fetal tissue are associated with a small but definite risk to the pregnancy even in experienced hands, so they are usually employed only when there is a high probability of a fetus being affected (e.g. because the condition has occurred in a previous child of the couple).

Increasingly, as the genes responsible for inherited metabolic diseases are identified, molecular genetic analysis is being employed for antenatal diagnosis. Indeed, it is now becoming possible to screen embryos obtained by in vitro fertilisation before implantation.

The principle of antenatal diagnosis by molecular genetic techniques is to screen fetal tissue directly for the mutation(s) in question. Fetal tissue can be obtained from 10 weeks of gestation by chorionic villus sampling. If the sequence of bases in the vicinity of the mutation is known, one of several modifications of the PCR (polymerase chain reaction, see p. 11) can be used to identify the presence or otherwise of the mutant gene.

Antenatal screening by detection of the mutant gene is best ilustrated by a consideration of cystic fibrosis. This autosomally recessively inherited condition results in greatly increased viscosity of exocrine secretions. Clinically, children usually present in the first year of life with recurrent respiratory infections (related to impaired clearance of bronchial secretions) and/or malabsorption due to pancreatic insufficiency. Presentation can also occur in the neonatal period with intestinal obstruction (meconium ileus) or cholestasis. Although, with modern treatment, children with cystic fibrosis usually survive into adulthood, this is not achieved without considerable cost to them and their carers (e.g. vigorous chest physiotherapy several times daily).

Cystic fibrosis is due to defective synthesis of a protein involved in transmembrane chloride transport (the 'cystic fibrosis transmembrane conductance regulator'). In 75% of cases, cystic fibrosis is due to a 3 bp deletion (F508). PCR techniques have been developed which allow simultaneous detection of this and the other common mutations. As a result, it is possible to screen for cystic fibrosis antenatally in families in whom the condition is known to occur, although cases will still be missed because a small number are caused by individually rare but collectively numerous 'private' mutations which only occur in single or small numbers of families.

Similar techniques can also be used to screen prospective parents for cystic fibrosis, as discussed in the next section.

7.4.3 *Screening for carriers*

When a kindred is known to be affected by a recessively inherited metabolic disease, the identification of carriers (heterozygotes) may be of value in genetic counselling— that is, on advising individuals on the risk of the disease occurring in their children.

The methods used can be divided into those based on the detection of phenotypic abnormalities, and those based on the detection of the abnormal gene. For recessively inherited conditions (which includes the majority of inherited metabolic diseases), heterozygotes are, by definition, clinically normal. Nevertheless, it may be possible to detect an abnormality in these individuals, for example by stressing an enzyme pathway by giving a loading dose of a substrate.

When (as is increasingly becoming the case) the base sequence of the abnormal gene is known, it should be possible to synthesize probes which can be used to detect the mutation directly. However, when the disease in question can be caused by more than one mutation, it would be necessary to be able to screen for every mutation to be certain of excluding the possibility that an individual was a carrier. While most cases of the condition may be due to one or a small number of mutations, the existence of private mutations (see above) inevitably reduces the reliability of screening. Currently, testing for the four commonest mutations allows unequivocal identification of approximately three-quarters of couples at risk of having a child with cystic fibrosis.

Screening methods have also been developed for use when the precise site of a mutation is unknown. Even when the gene causing a disease has not been identified, it may be possible to detect heterozygotes if a genetic marker can be identified which is closely linked to the gene—i.e. the marker is always present when the mutant gene is present.

Polymorphisms for sites susceptible to cleavage by restriction endonucleases (restriction fragment length polymorphism, RFLP) have been widely used for this purpose.

These techniques are also applicable to antenatal screening of fetal samples. The interested reader should consult the works cited in 'Further reading' at the end of this chapter for more details of these techniques.

7.5 Treatment of inherited metabolic disease

There are many possible approaches to the treatment of inherited metabolic disease; these are summarised in Box 7.2.

7.5.1 *Gene therapy*

Definitive treatment would involve gene therapy—i.e. replacement or modification of the defective gene to permit normal expression of its product. While this approach was considered by many to be in the realms of fantasy even within the present decade, it is now the focus of considerable research and some clinical trials have been performed while others are in progress. Among the problems that need to be overcome are not just development of replacement genetic material, but how to introduce it into the genome of cells in a way that will ensure its replication when the cell divides, how to ensure that it is subject to normal control of its expression and how to avoid its incorporation into germ cell lines which might result in its transmission to any chil-

Box 7.2 **Potential treatments for inherited metabolic diseases**

- Gene therapy
- Treatment with product of defective gene
- Enhancement of enzyme activity
- Treatment with product of defective enzyme
- Prevention of accumulation of toxic metabolite
- Removal of toxic metabolite
- Avoidance of factors precipitating acute illness

dren, in whom the consequences of its presence would be entirely unpredictable. Both ex vivo and in vivo gene delivery techniques have been tried. In the former, retroviruses have been used to transfect genes into cells taken from the patient which are then returned to them. In vivo methods use an adenovirus as a vector for the gene or involve giving DNA–liposome complexes. To date, the condition most targeted for gene therapy has been cystic fibrosis. Early results, whilst not coming anywhere near achieving a cure, are nonetheless are encouraging. This subject is in its infancy and there seems little doubt that this method of treating inherited metabolic disease has an exciting future.

When an inherited disease is manifest as a result of a defect in an enzyme in a single organ, organ transplantation, which effectively replaces the defective gene in the critical organ, may be an option. This has been used successfully in patients with homozygous familial hypercholesterolaemia (liver transplantation) and cystinuria (kidney transplantation) and in a variety of lysosomal storage disorders (bone marrow transplantation). Despite the increasing success of such procedures, transplantation may not be considered an appropriate technique if it is feasible for the condition to be adequately treated by other means. There are many facets to a consideration, for example, of the relative advantages and disadvantages of organ transplantation and conventional treatment. Thus the former may 'cure' an individual with a metabolic disease (but requires the taking of lifelong immunosuppressive therapy), whereas conventional treatment may impose a considerable burden on the quality of life by, for example, requiring strict compliance with an unpalatable diet.

7.5.2 *Replacement of the missing gene product*

This approach to the management of inherited metabolic diseases has the attraction that it does not involve genetic manipulations but the disadvantage that the benefits of such treatment, unless it is continually repeated, are short-lived because of the normal turnover of all body tissues.

This technique is the standard approach to the management of inherited disorders of blood coagulation (e.g. haemophilia, due to defective synthesis of coagulation factor VIII) and to some humoral immunodeficiency disorders, where immunoglobulin synthesis is defective.

The situation is more difficult where the defective protein is intracellular. Attempts have been made to treat lysosomal storage diseases (conditions characterised by the accumulation of macromolecules which are normally broken down by lysosomal enzymes) by incorporating the normal enzyme into lipid particles ('liposomes') and injecting them into the circulation, whereupon they are phagocytosed by cells and incorporated into lysosomes, but with little success.

7.5.3 *Enhancement of the activity of a missing enzyme*

This approach has been used with success for a limited number of conditions in which the defective enzyme is activated by a vitamin-derived cofactor. It cannot be effective if the mutation results in a complete absence of enzyme synthesis. Thus, some patients with methylmalonic aciduria, a group of conditions characterised by defective carboxylation of propionyl-CoA to form succinyl-CoA, respond to high doses of hydroxycobalamin (vitamin B_{12}), a derivative of which is a cofactor for the enzyme responsible, methylmalonyl-CoA mutase. Similarly, some patients with primary hyperoxaluria type 1 (glycolic aciduria) respond to pharmacological doses of pyridoxine (vitamin B_6), the coenzyme for the defective enzyme, glyoxylate:alanine aminotransferase. Other examples are given in Table 7.1.

Table 7.1 Examples of treatment of inherited metabolic disease by vitamin supplementation

Disease	Vitamin
Methylmalonic aciduria (propionyl-CoA carboxylase deficiency)	B_{12}
Maple syrup urine disease (branched-chain keto acid dehydrogenase deficiency)	Thiamin
Glycolic aciduria (Glyoxylate:alanine aminotransferase deficiency)	Pyridoxine
Homocystinuria (cystathionine β-synthase deficiency)	Pyridoxine
Cystathioninuria (γ-cystathioninase deficiency)	Pyridoxine
Vitamin D-dependent rickets (renal 1α-hydroxylase deficiency)	Cholecalciferol*

*Or synthetic 1-hydroxylated derivatives.

7.5.4 *Replacement of the missing product of an enzyme*

In those inherited metabolic diseases where some or all of the consequences relate to decreased synthesis of the product of the defective enzyme, provision of that product, where possible, would be expected to improve the outcome. This approach to treatment can be illustrated by the syndromes of congenital adrenal hyperplasia. As discussed above (p. 136) the clinical features in the commonest type derive from the decreased synthesis of cortisol (and sometimes aldosterone) and increased synthesis of adrenal androgens. Treatment with cortisol reverses the consequences of lack of this hormone, including the increased synthesis of pituitary ACTH which drives the increased androgen production. Treatment with a mineralocorticoid prevents the sodium-wasting and retention of potassium that occurs if aldosterone synthesis is also affected.

The fasting hypoglycaemia (although not all the metabolic sequelae) of glycogen storage disease type 1 (glucose 6-phosphatase deficiency) can be treated by giving affected children frequent meals containing corn starch, which is only slowly broken down into glucose in the gut, ensuring a constant supply into the bloodstream. Overnight nasogastric tube feeding may be required in addition, especially in small infants prone to frequent hypoglycaemia.

7.5.5 *Measures to decrease the accumulation of a toxic substrate*

The clinical features of many inherited metabolic diseases result from the accumulation of a toxic precursor of the defective enzyme (p. 131). The most usual method to decrease such accumulation involves modification of the diet in order to reduce supply of substrate to the enzyme. An obvious example is phenylketonuria; restriction of dietary phenylalanine decreases the plasma concentration of this essential but potentially toxic amino acid and, if treatment is started shortly after birth, the otherwise irreversible impairment of mental function that occurs in this condition can be prevented. Because phenylalanine is an essential amino acid, the diet must not be totally lacking in it; because the formation of tyrosine from phenylalanine is defective in this condition, supplementation of the diet with tyrosine is required. The diet is unpalatable, and frequent monitoring of affected children is required to ensure that plasma concentrations of the amino acid are adequate for normal protein synthesis but remain below the levels that cause neurotoxicity. The diet often has to be relaxed somewhat (with consequent increases in plasma phenylalanine concentrations) in later childhood in order to achieve compliance, but it should not be abandoned. There is increasing evidence that doing so leads to a poorer outcome. Strict diet is necessary should a woman with phenylketonuria become pregnant, in order to protect the fetus from the damaging effects of exposure to high maternal concentrations of phenylalanine.

Among many other disorders that respond to dietary modification are galactosaemia (avoidance of dietary galactose), hereditary fructose intolerance (avoidance of dietary fructose, see p. 150) and some urea cycle disorders (decreasing dietary protein content which limits ammonia accumulation).

7.5.6 *Removal of toxic substrates and metabolites*

This is another well-established means for treating certain inherited metabolic diseases. Wilson's disease is a disorder of copper metabolism, in which there is decreased protein binding of copper in the blood and accumulation of the metal in many tissues, including the liver, central nervous system and kidneys. The condition can be effectively treated by the frequent administration of penicillamine, which chelates copper and promotes its urinary excretion. Primary genetic haemochromatosis, in which increased absorption of dietary iron leads to iron overload with damage to the liver, endocrine organs and heart, can be treated by regular venesection to remove iron in haemoglobin. Haemopoiesis is stimulated, with increased mobilisation of tissue iron for incorporation into haemoglobin. Treatment with the iron-chelating agent, desferrioxamine, is much less effective because, even though plasma iron concentrations are increased in haemochromatosis, most of the iron is protein-bound, and not accessible to the chelator.

Severe cases of familial hypercholesterolaemia have been treated successfully with a technique called LDL (low-density lipoprotein) apheresis, in which LDL particles are physically removed from the plasma using a binding resin or specific antibodies in an extracorporeal circuit.

In all instances where inherited metabolic diseases are susceptible to treatment to remove toxic substances, treatment must either be continuous, or repeated frequently, in order to prevent a resurgence of the clinical features.

7.5.7 *Avoidance of precipitating factors*

Some inherited metabolic diseases may only become apparent if a precipitating factor is present. This is frequently dietary. An example is provided by the condition hereditary fructose intolerance. This is due to a partial defect in the enzyme fructose 1-phosphate aldolase which leads to accumulation of fructose 1-phosphate when fructose is present in the diet (Figure 7.7). This in turn inteferes with glucose homoeostasis and causes hypoglycaemia. The condition is clinically silent until infants are exposed to fructose (or sucrose, which contains fructose), and avoidance of this sugar prevents the manifestations of the enzyme defect. Recognition that a specific pattern of clinical features follows a change in diet may provide an important clue to the diagnosis of inherited metabolic disease.

Precipitating factors other than diet can also be important. Perhaps the best examples in this category are provided by the porphyrias (p. 134). Clinical attacks of the primarily neurological porphyrias are often precipitated by drugs which may act by increasing the requirement for haem, thus decreasing the negative feedback on the rate-limiting step and increasing the synthesis of the neurotoxic porphyrin

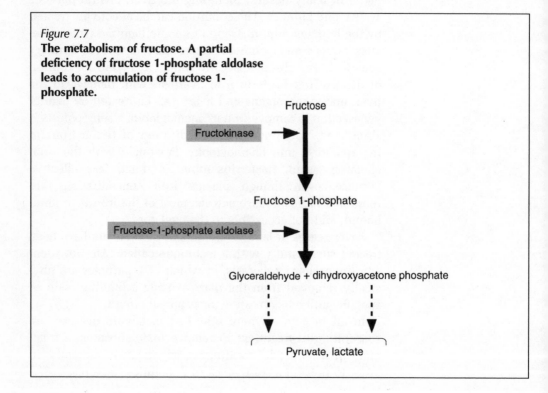

Figure 7.7
The metabolism of fructose. A partial deficiency of fructose 1-phosphate aldolase leads to accumulation of fructose 1-phosphate.

Fructose

Fructokinase →

Fructose 1-phosphate

Fructose-1-phosphate aldolase →

Glyceraldehyde + dihydroxyacetone phosphate

Pyruvate, lactate

precursors. With the chronic (photosensitising) porphyrias, prevention of exposure of the skin to ultraviolet light (e.g. with barrier creams) is the mainstay of treatment.

7.6 Conclusion

The inherited metabolic diseases include thousands of diverse conditions. All are uncommon, although collectively they are a major cause of illness, particularly in childhood. Their effects vary between being universally fatal early in life and being clinically silent and discovered as incidental findings. Effective treatment is available for some, and increasing understanding of their genetic basis and of techniques for genetic manipulation offers exciting possibilities for advances in treatment in the future.

Further reading

Alper, J.S., 1996. Genetic complexity in single gene disease. *British Medical Journal*, **312**, 196–197. (A short review explaining some of the reasons for clinical diversity within individual inherited diseases.)

Dorin, J., 1996. Somatic gene therapy. *British Medical Journal*, **312**, 323–324. Leide, J.M., 1995. Gene therapy—promise, pitfalls and prognosis. *New England Journal of Medicine*, **333**, 871–873. (Two leading articles which give up-to-date summaries of the present status of this form of treatment.)

Holton, J.B., 1994. *The Inherited Metabolic Diseases*, 2nd edn. Edinburgh: Churchill Livingstone. (A concise account of the major groups of these conditions.)

Scriver, C.R., Beaudet, A.L., Sly, W. and Valle, D. (Eds), 1995. *The Metabolic and Molecular Basis of Inherited Disease*, 7th edn. New York: McGraw-Hill. (The definitive account of the subject, in three volumes.)

Trent, R.J., 1993. *Molecular Medicine*. Edinburgh: Churchill Livingstone. (Aptly subtitled 'An Introductory Text for Students', this book provides an exceptionally clear account of the principles and applications of molecular genetics.)

8 Disease and Genetic Polymorphism

Michel R. A. Lalloz

8.1 Genetic disease

Hippocrates, the father of Western medicine, was aware that individual differences (polymorphism), such as eye colour, baldness and certain illnesses, could be passed on from parents to their children. As early as the first few centuries AD, the Jewish Talmud recognised that if two male infants within one family had bled to death following circumcision (due to the condition now known as haemophilia) then subsequent boys should be exempt from circumcision.

Transmission genetics, established in 1865 by Mendel, gave us the notion of allelic segregation, where paired chromosomes do not normally enter the same gamete, and of independent assortment, where alleles at different loci assort at random during meiosis. He also identified paired units of inheritance, namely genes. Also in 1865, Galton founded biometric genetics, and statistical methods were applied to inherited human characteristics, which subsequently formed the basis of population genetics. The discovery by Landsteiner of the ABO blood group polymorphism in 1900, which was subsequently elucidated 90 years later when Yamamoto showed that the different blood types are caused by multiple alleles at a single locus, illustrated the notion of non-detrimental inherited variation between individuals.

It was Garrod in 1902 who first associated a well-characterised human condition, alkaptonuria, with a genetic defect resulting in the deficiency of an enzyme. In 1909, Garrod published a classic book, 'Inborn Errors of Metabolism', in which he recognised that biochemical defects were hereditary and occurred in siblings from normal parents. Human gene mapping was initiated in 1911 by E.B. Wilson, who deduced that the gene for colour blindness is located on the X chromosome. In the 1930s, the association made between colour blindness and haemophilia (i.e. that males with haemophilia in the family were also colour blind) led to the concept of genetic linkage.

In 1949, sickle cell anaemia was identified to be hereditary (Neel) and shown to be due to a chemical modification in haemoglobin detectable by altered electrophoretic mobility (Pauling). This condition became the first inherited disease to be characterised at the molecular level when, in 1957, Ingram showed the cause to be due to a substitution of valine for glutamic acid at amino acid position 6 in β-globin. The region of the genome encoding β-globin subsequently became the first human gene to be cloned and sequenced and the specific sickle defect shown to be due to an adenine-to-thymine base change (see p. 167). Human genetics has since been transformed into a discipline, dominated by molecular explanations of disease, aided by rapid technological advances facilitating a detailed molecular analysis of man's genetic make-up.

8.2 The human genome

The human genome consists of tightly coiled molecules of deoxyribonucleic acid (DNA) organised into chromosomes. DNA encodes all the information necessary for building and maintaining life. Four bases, adenine (A), cytosine (C), guanine (G) and thymine (T), are present in DNA. The order of these nucleotides along the DNA strand define the DNA sequence and this sequence specifies the precise genetic constitution of the individual. Base-pairing rules dictate that cytosine invariably pairs with guanine (C–G) and adenine with thymine (A–T). Weak hydrogen bonds between the complementary base pairs (bp) bind the two DNA strands together to form a double helix.

The entire genome of about 3×10^9 bp is duplicated each time a cell divides (mitosis or meiosis). The DNA molecule in the nucleus of the cell unwinds and the strands separate. Free nucleotides pair up with their complementary bases on the strand such that each DNA molecule synthesises (replicates) a complementary new strand. One new and one old DNA strand are inherited by each daughter cell. Complementary base-pairing minimises errors (mutations) which may affect the individual (somatic mutation) or their progeny (germ line mutation).

A specific sequence of nucleotides encoding the information required for constructing a protein is a gene. Genes are the basic physical and functional units of heredity. The human genome comprises at least 100 000 genes. Although genes often span thousands of bases (kb), less than 10% of the genome constitutes coding sequences (exons). In many genes, these are separated by non-coding sequences (introns). Non-coding intergenic regions comprise control sequences and other regions whose functions remain obscure. Within each exon three consecutive bases (codon) codes for one of the 20 naturally occurring amino acids (e.g. codon ACG codes for the amino acid threonine). Thus, codon order specifies the sequential arrangement of amino acids within proteins.

Genes are often under the control of upstream (5′) regulatory elements constituting defined sequences such as TATAAA and CCAAT motifs which play a role in expression, enhancers and repressors involved in induction and regulation, and other responsive sequences which confer tissue specificity. These motifs (*cis*-elements) bind proteins (*trans*-acting DNA-binding proteins) required for appropriate regulation of gene function. Downstream (3′) of the gene are found specific sequences involved in the accurate processing of genetic information.

Protein-coding instructions from the gene are expressed as messenger ribonucleic acid (mRNA). This is a complementary RNA strand synthesised (transcribed) from the DNA template in the nucleus and which passes into the cellular cytoplasm where it in turn serves as the template for protein synthesis. Nuclear pre-mRNA encoding a two exon gene is processed by first cleaving the 5′ (donor) splice site generating an exon-containing RNA species and a lariat RNA containing the intron and second exon. Cleavage at the 3′ (acceptor) splice site and ligation of the exons results in excision of the intron in the form of a lariat. The accuracy of cleavage and rejoining RNA is determined by consensus sequences spanning 5′ and 3′ splice junctions which include the invariant GT and AG dinucleotides present at the 5′ and 3′ exon/intron junctions, respectively. Cytosolic mRNA is then translated into a string of amino acids constituting the protein for which it codes.

The human genome is organised into chromosomes along each of which genes are arranged sequentially. Somatic cells contain two sets of chromosomes, one set inherited from each parent. Each set has 22 autosomes and a sex chromosome; a female will have two X chromosomes and a male an X and a Y chromosome. Chromosome staining can, under a light microscope, reveal light and dark banding patterns reflecting variation in G/C or A/T DNA content. Size and banding pattern differences allow chromosomes to be distinguished from each other by karyotype analysis. Lack or additional presence of a chromosome or a gross major break and rejoining (translocation) are detectable by this method. More subtle changes in DNA require detailed molecular analysis. Such changes (mutations) are responsible for most inherited diseases, including sickle cell anaemia, haemophilia and cystic fibrosis, or may predispose to cancer, psychiatric illness and other complex diseases.

8.3 DNA polymorphism

In essence, genetic polymorphism is the difference in DNA sequence between individuals. Variations in the human genome will arise from exogenous factors such as exposure to chemical mutagens or ionising ultraviolet light irradiation. Others may arise from error-prone in vivo processes such as DNA replication and repair. These mechanisms can be enzymatic such as postreplication mismatch repair or exonucleolytic proofreading, physical as in DNA slippage, or chemical such as deamination of 5-methylcytosine. If a polymorphism occurs within a functional region of a gene, regulatory, coding or processing sequence, then the potential for a deleterious effect is possible. If a polymorphism occurs elsewhere then a potentially useful marker may have been created.

Polymorphisms in DNA sequence occur on average once every 300–500 bp. Polymorphisms within exons may lead to observable changes that need not be detrimental, such as differences in blood group (see p. 153). The redundancy built in to the genetic code will often result in silent single-nucleotide changes. A codon change in the third base from CTA to CTG or in the first base to TTA is silent since all three triplets code for the amino acid leucine. Some mutations may result in subtle changes to protein sequence. The GGG-to-GCG mutation in the middle base of the codon results in a conservative amino acid substitution, glycine to alanine, since the difference between the physical and chemical properties of the two amino acids is minor. In other instances the substituted amino acid may have little or no effect on the normal function of the protein.

Most mutations which give rise to polymorphisms occur within introns and between genes and have little or no effect on an organism's appearance or function. Some, however, can have marked effects if they occur within cis-elements which affect gene expression (see p. 155), or mRNA processing sequences. The latter include abolition of consensus splice site recognition sequences, the creation within introns of cryptic splice sites, and alteration to the polyadenylation, poly(A)$^+$, signal which confers stability to the mRNA.

All polymorphisms are detectable at the DNA level. A polymorphism detectable in one individual or a single family only is likely to be of recent occurrence and is referred to as a private polymorphism. Polymorphisms that occur in a large population and are present in a wide range of ethnic groups will have arisen some considerable time ago.

Examples of markers that are easily detectable in the laboratory are restriction fragment length polymorphisms (RFLPs), with DNA sequence variations at a single base that can be cleaved by DNA restriction enzymes, and short tandem repeat polymorphisms (STRPs), which vary in the number of repeated units and, therefore, in sequence length.

Restriction enzymes recognise specific DNA sequences and cut the molecule at those sites. Some enzymes cut DNA infrequently (rare-cutters), generating large fragments (several thousand bp). Most enzymes cut DNA more frequently, generating many small fragments (<100 to >1000 bp). Restriction enzymes with four-base recognition sites (four-cutters) will yield pieces averaging 250 bp long, six-cutters yield average pieces of 4000 bp and eight-cutters yield pieces of about 64 000 bp. A single base change can either create a sequence recognised by an enzyme or abolish such an already existing site. Since a restriction site is either present or absent, RFLPs are bi-allelic and therefore have limited information content.

STRPs (also referred to as microsatellite markers and variable number tandem repeats) are widely dispersed in the human genome. Their ubiquitous presence, ease of typability by the polymerase chain reaction (PCR) method, Mendelian co-dominant inheritance and multi-allelic polymorphic character has led to the development of genome-wide genetic maps utilising these as markers (see p. 159).

Little is known about the nature of spontaneous mutation in eukaryotes, let alone man. Point mutations in human genes, however, are non-random. Methylated CpG sites are 'hot spots' for mutations as evidenced, first, by the high frequency (33%) of CG to CA and CG to TG transitions among mutations causing human genetic disease, and, secondly, the high rate of polymorphism detected by restriction enzymes (see p. 165) with CpG within their DNA recognition sequence. Deamination of the CpG dinucleotide 5-methylcytosine to give either a G to A or a C to T substitution, depending upon which strand the 5-methylcytosine is mutated, is one well-characterised mechanism of in vivo mutagenesis.

Insertion or deletion mutations of a single or many bases within a coding sequence usually have notable detrimental consequences on gene transcription. A point insertion or deletion within a coding sequence causes a single base shift in the sequence reading frame (frameshift), resulting in incorrect transcription and subsequent translation of the gene. Such mutations can create a downstream premature termination codon which often results in the transcription of unstable mRNA. Regions of single base repeats such as TTTTTT are vulnerable to point mutation, which can result by slippage or looping out of a base and the repair mechanisms failing to identify and correct the errors.

A trinucleotide microsatellite exhibiting high variability in number of repeats is responsible for gene instability and the molecular pathology of a number of inherited diseases. The simple $p(CCG)_n$ repeat is present within the 5' region of the *FMR-1* gene located at Xq27.3. Expansion of this repeat abolishes gene expression, causing fragile X syndrome in affected males. Normal individuals have $(CCG)_{6-60}$ repeats in the *FMR-1* gene. Carrier males capable of transmitting the condition have $(CCG)_{60-200}$ copies and affected subjects have $(CCG)_{>200}$ repeats. Amplification of a repeated $p(CTG)_n$ trinucleotide DNA sequence that lies, in contrast to the *FMR-1* gene, within the 3' untranslated region of a transcript (mapped to 19q) homologous to the protein kinase gene family gives rise to myotonic dystrophy (DM). Normal individuals have $(CTG)_{5-37}$ whereas DM patients have $(CTG)_{50}$ to $(CTG)_{>1000}$. The length of the (CTG) repeat correlates with the DM phenotype. Moreover, (CTG) repeat instability extends beyond meiotic alteration to mitotic changes manifest as somatic variation. Thus, differences in (CTG) copy number need not only be vertical within a pedigree but also horizontal.

During meiosis (see above) DNA strands break and can rejoin either interchromosomally, on the other copy of the homologous chromosome, or intrachromosomally, in a different location on itself. Homologous recombinations are a known cause of mutations within the α-globin gene cluster (gene deletion and insertion). Intrachromosomal inversion as a cause of recurrent mutation due to homologous recombination has been reported for only one gene. Within intron 22 of the factor VIII gene is entirely contained the intronless gene *F8A*, which is transcribed in the reverse orientation to factor VIII. This *F8A* gene can recombine with either of two copies of itself (500 kb upstream of factor VIII) transcribed in the same orientation as factor VIII.

8.4 Detecting genes that cause genetic diseases

The difficulty in finding a disease gene of interest depends on what information is known about the abnormality, especially DNA alterations which can cause the disease. Identifying the disease gene is difficult when disease results from a single altered DNA base as is likely the case for most major human inherited diseases. Positional cloning can identify the gene responsible for a genetic disease (Figure 8.1). Genetic and physical map information (see below) can combine to locate a new gene to a genomic region. This approach is used when no cytogenetic rearrangement or biochemical defect is identified, and no animal model for the disease exists.

Identification of all human genes will expedite the efficiency with which disease-causing and disease-susceptible genes are characterised. Means by which to utilise this information will be developed to further the study of biology and medicine. To achieve this, however, chromosomes have first to be divided into fragments that can easily be studied, these fragments can then be ordered with respect to their relative location (mapping) on each chromosome. The base sequence of each ordered DNA fragment will ultimately complete the detailed map of the human genome.

A genetic linkage map shows the order of specific DNA markers along a chromosome. Markers include genes as well as DNA segments with no known function but whose inheritance pattern can be followed through the generations. Such markers must, therefore, exist as alternative forms among individuals so as to be distinguishable between different members in family studies. Collaboration between the Cooperative Human Linkage Center (CHLC), Généthon, the University of Utah Human Genome Center, and the Centre d'Étude du Polymorphisme Humain (CEPH), has resulted in an integrated human map incorporating highly polymorphic STRP markers consisting of di-, tri- and tetranucleotide repeats.

The frequency with which two markers are inherited together allows the construction of the human genetic linkage map. Markers close to each other on a chromosome will more often be inherited together. During meiosis, DNA strands can recombine either interchromosomally or intrachromosomally (see above). Meiotic recombination can thus separate two markers originally located on the same chromosome. Markers near each other are tightly linked and less likely to recombine. Recombination frequency thus provides an estimate of the distance separating markers. Markers 1 centimorgan (cM; named after Thomas Hunt Morgan) apart are separated by recombination 1% of the time. A genetic distance of 1 cM approximates a physical distance of 1×10^6 bp (1 Mb).

A putative inherited disease gene can thus be located by tracking the inheritance of polymorphic DNA markers in affected pedigrees. Genetic maps have allowed the exact chromosomal location of several disease genes to be identified, including cystic fibrosis, fragile X syndrome, myotonic dystrophy and Tay–Sachs disease.

Physical maps are briefly outlined. The basis of the chromosomal map is the banding patterns observed by light microscopy of stained chromosomes (see p. 155). DNA fragments including genes are assigned to specific chromosomal locations and distances measured in bp. Fluorescence in situ hybridisation (a technique utilising a fluorescence labelled DNA marker) allows DNA sequences as close as 2–5 Mb to be oriented on a chromosome. Modified in situ hybridisation procedures using chromosomes at interphase, when they are less condensed, increase map resolution to around 100 000 bp. A complementary DNA (cDNA) map locates exons on the chromosomal map. cDNA synthesised from isolated mRNA by reverse transcription can be mapped to genomic regions. A cDNA map thus locates genes with unknown functions to specific chromosomal regions. These expressed sequence tags may represent parts of the genome with the most biological significance. Such a map can clearly suggest candidate genes to test when a disease gene has been located to a region of the genome by genetic linkage techniques. A cosmid contig maps the overlapping DNA fragments spanning the genome. A macrorestriction map shows the order and distance between enzyme recognition sites. Ultimately, the high-resolution physical map aimed at is the DNA base-pair sequence of each chromosome in the human genome.

Positional cloning is achieved in several stages (Figure 8.1). First, the region containing the defective gene is localised. This requires DNA from the families of patients with the disease and polymorphic markers selected to cover the area of the genome to be investigated, the aim being to identify the markers which specifically segregate with the disease. This should define a DNA interval containing the disease gene flanked by markers from the genetic linkage map. Mapping disease phenotypes for autosomal dominant disorders where large pedigrees are available remains

Figure 8.1
Positional cloning may involve a genome-wide search for the gene responsible for the disease. Identifying the markers that specifically segregate with the disease will define a DNA interval within which genes can be expressed and analysed for the presence of a disease-causing mutation.

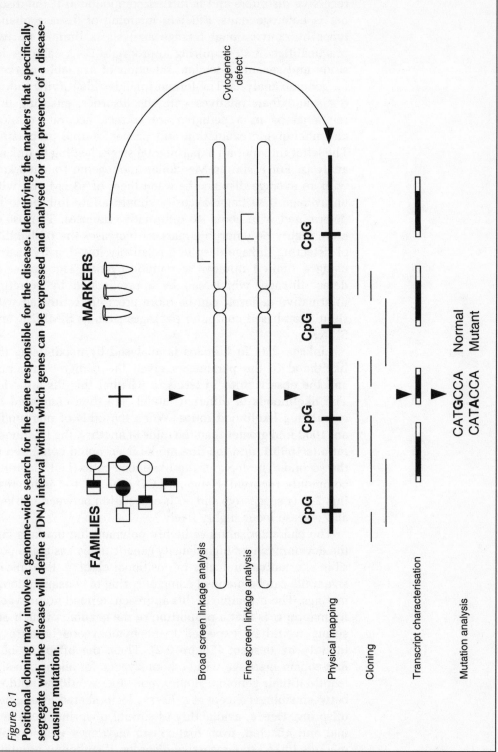

labour-intensive. The order of difficulty increases for rare recessive disorders and is further compounded if the disorder is heterogeneous. Efficient mapping of disease phenotypes by conventional linkage analysis is limited by two practicalities, first, acquiring appropriate DNA samples for study and second, adequate selection of available markers for genome analysis. The former includes identifying individuals and their relatives with the disorder, ensuring that relationships in a pedigree are correct, accurate clinical and phenotypic evaluation and proper sample processing. The latter includes all experimental stages leading to linkage analysis. For a simple Mendelian monogenic trait markers with an average distance between them of 13 cM and with on average 0.80 heterozygosity should allow linkage to be determined with about 40 informative meioses. The use of more highly polymorphic markers increases the probability of detecting linkage despite a relatively low-density search using a limited number of meiotic events. The power to detect linkage, which can be approximated by counting informative meioses, can be more precisely estimated with simulation-based computer packages such as SIMLINK and SLINK.

Linkage data in humans is analysed by maximising the likelihood of the parameters given the pedigree structure and the observations in terms of whether this fits a model. The likelihoods of different models are then compared by calculating likelihood ratios. When the odds of one model are 1000 fold greater than the odds of another, the likelihood ratio is 1000:1, and the first model is accepted compared to the second. The $\log_{10}$ of the likelihood ratio (LOD score) is commonly reported. Hence, a LOD score of 1–2 is interesting, 2–3 is suggestive and >3 is association between markers and disease locus highly likely.

The characterisation of highly polymorphic markers and the development of high-density genetic maps has made possible alternative strategies to positional cloning. Rare recessive traits can be mapped using offspring of consanguineous matings. The principle of this approach, termed homozygosity mapping, is that a proportion of the genome of such offspring would be expected to be homozygous because of identity by descent (Figure 8.2). Thus, the offspring of a first-cousin marriage will be homozygous for about one-sixteenth of their genome. Homozygous loci would be random between siblings except at a disease locus shared by affected offspring. Hence, availability of several offspring, both normal and affected, from first-cousin marriages will identify markers linked to a recessive disorder. If sufficient numbers

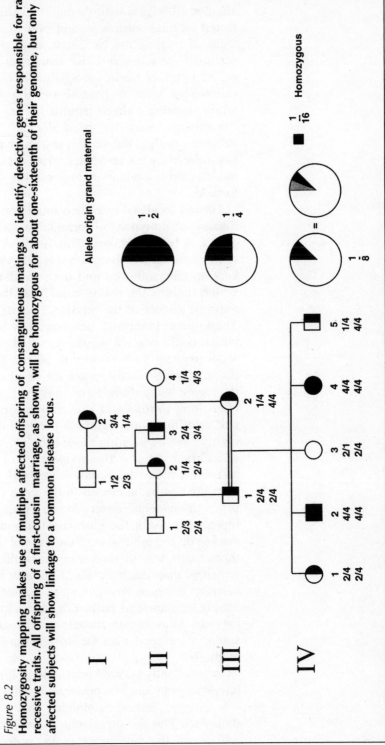

Figure 8.2
Homozygosity mapping makes use of multiple affected offspring of consanguineous matings to identify defective genes responsible for rare recessive traits. All offspring of a first-cousin marriage, as shown, will be homozygous for about one-sixteenth of their genome, but only affected subjects will show linkage to a common disease locus.

of offspring are available then equimolar concentrations of genomic DNA can be pooled such that samples from all affected siblings constitute one pool and samples from unaffected siblings form a second pool. The two major advantages of pooling are that, first, only two samples need be screened for any one STRP marker at a time and, second, gel analysis of allele segregation is inspected visually for differences between control and affected pools, the latter when showing a single band being suggestive of linkage to the disease locus. The more distantly related the affected subjects are then the smaller the shared segment containing the defective gene is. Such a disease-causing gene can only be detected if a relatively high marker density screen is performed.

Having localised a chromosomal region associated with a disease phenotype, the formidable task of identifying the gene can be undertaken. This involves isolating the region of the genome which covers this interval. The flanking markers are characterised and used to identify clones of DNA which include the region containing the disease gene. This requires the use of the physical map of the human genome. The interval identified, however, may be as large as 10 Mb, which could contain several hundred genes. Thus, the next stage requires an inventory of all the genes present within the interval. Since the genes are small relative to the size of the region being studied and they are divided into exons and often long introns and also they are randomly distributed and vast intergenic regions separate them, this represents a highly time-consuming procedure. Once a marker is linked to within 1 cM of a disease gene, chromosome walking can be used to clone the disease gene itself. A genomic fragment identified as being closest to the gene of interest is used to probe the genomic library for an overlapping clone. This is repeated to walk the chromosome and reach the flanking marker on the other side of the gene of interest. In principle, there exists within such a region sufficient information by which to map the intervals of identity by descent and non-identity between two genomes. Unfortunately, few of the couple of thousand base differences found per cM (0.1%) between homologous chromosomal regions from unrelated subjects are accessible for linkage studies by conventional methods.

In genetically isolated populations, linkage disequilibrium (allele association) is a powerful strategy for identifying disease-causing genes originating in a common ancestor (founder). The defective gene in multiple affected subjects will retain the pattern of alleles present on the ancestral chromosome, the vicinity of the disease gene showing greatest retention. This approach exploits numerous historical meioses resulting in higher recombinational resolution. Such mapping applied to the Finns (founded about 100 generations ago) allowed cloning of the *DTD* gene. Conventional recombinant mapping localised the gene to an interval of 1.5 cM, whereas linkage disequilibrium considerably narrowed it down to an interval of 50 kb.

The final stage in positional cloning is identification of the mutated gene. Having characterised all the genes in the region of interest, one of them must be identified as defective for the disease being studied. Comparing the gene in diseased patients with the gene in healthy individuals should identify a mutation which segregates specifically in affected subjects within a kindred. Isolation of the gene will allow its function to be studied and its role in the development of disease elucidated. This should expedite effective therapeutic strategies such as pharmacology and gene therapy.

8.5 Methods used to detect mutations

Cloning and sequencing, although capable of detecting any mutation, is not a rapid method of mutation detection for even moderately sized genes. Selective methods such as Southern blotting of genomic DNA after digesting with *Taq*I is limited because only mutations within the TCGA recognition sequence of the enzyme (i.e. CpG alterations) will be detected. In general, screening methods capable of detecting any mutation is desirable.

The advent of the PCR method has greatly increased the scope and range of methods available for mutation hunting. Chemical cleavage of mismatches, denaturing gradient gel electrophoresis (DGGE) and single-stranded conformation polymorphism (SSCP) analysis have all been successfully applied to mutation hunting in a wide range of genes (Figure 8.3). Chemical cleavage and DGGE are both dependent upon the formation of homo- and heteroduplexes between normal strands of PCR-amplified DNA and a similarly amplified complementary strand containing a mismatch (Figure 8.3a). Chemical cleavage depends upon piperidine cleavage at the mismatch in a heteroduplex after chemical modification by hydroxylamine or osmium tetroxide (Figure 8.3b). In contrast, electrophoresis in a denaturing gradient gel, put simply, results in the mutation being detected when the heteroduplex ceases to migrate due to strand separation at a point different to the homoduplex

Figure 8.3

Four different approaches to DNA mutation analysis. (a) DGGE requires special equipment, computerised optimisation and costly GC-clamp primers. (b) Chemical cleavage is labour-intensive and uses high levels of radioactivity. (c) SSCP is relatively simple to set up and rapid to execute but may be less sensitive than chemical cleavage (b). (d) As for (b).

(a) **Denaturing gradient gel electrophoresis (DGGE)**

Normal	Mutant
NATGCTGGTAN	NATGCCGGTAN
NTACGACCATN	NTACGGCCATN

Heteroduplex (H)

T	C
NATGC GGTAN	NATGC GGTAN
NTACG CCATN	NTACG CCATN
G	A

(b) **Chemical cleavage of mismatches**

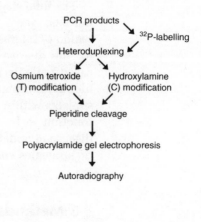

(c) **Single-strand conformation polymorphism (SSCP) analysis**

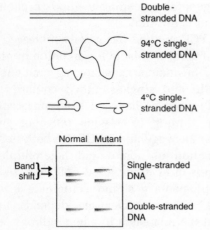

(d) **Ectopic transcript analysis by chemical cleavage**

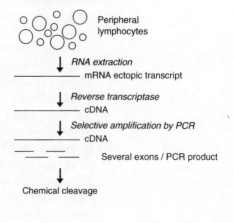

(Figure 8.3a). The principle of SSCP is that single DNA strands of different sequences exhibit different mobilities when electrophoresed in non-denaturing polyacrylamide gels such that a single base substitution can cause quite a different conformation and hence gel migration when compared to normal (Figure 8.3c). Small fragments are run directly on SSCP gels in order to optimise the mutation detection rate. Large fragments (>400 bp) for SSCP analysis are digested with an appropriate enzyme before electrophoresis. The three methods described all require PCR amplification of fragments, including intron/exon boundaries for each exon. Illegitimate transcription overcomes the problem of looking at individual exons by isolating RNA from peripheral lymphocytes and making cDNA by reverse transcription (Figure 8.3d). Defined regions of the cDNA can then be amplified with more exons within fewer fragments being simultaneously analysed by chemical cleavage.

Although chemical cleavage indicates the location of the alteration, it is clear that SSCP analysis, after restriction enzyme digestion, can also narrow down the region within which the alteration may be found. All the methods ultimately require identification of the specific alteration by DNA sequencing. While the methods described are clearly capable of detecting point mutations, other mutations, including large deletions, insertions, inversions and translocations, may require fine mapping using the Southern blot approach.

8.6 Carrier detection and prenatal diagnosis of inherited diseases

Mutation-specific analysis is clearly the most ideal approach to both carrier and prenatal diagnosis. This has been feasible for the condition of sickle cell anaemia since the A-to-T base change in codon 6 (GAG; Glu → GTG; Val) of the β-globin gene was found to abolish the restriction site for the endonucleases *Mst*II and *Cvn*I. Initially, diagnosis of sickle cell anaemia was achieved by Southern blotting and in situ hybridisation, but has since been superseded by the use of the amplification refractory mutation system or by *Mnl*I digestion (a restriction enzyme whose recognition site is also abolished by the mutation) of PCR-amplified product.

The specific gene defect responsible for an inherited disease within a given family is often unknown. If a pregnancy is involved, or one is planned imminently, and the gene exhibits considerable heterogeneity of molecular defects responsible for the disease, then mutation analysis is not the most expedient approach to diagnosis. In practice, the ability to identify the abnormal gene and thus trace its inheritance is all that is required of carrier detection and prenatal diagnosis. DNA polymorphisms can distinguish the two alleles of a gene regardless of the specific molecular defect. Such a polymorphism is independent of the cause of the disease but marks the normal and defective gene on the two chromosomes. Since DNA polymorphisms are abundant, characterisation of a newly cloned gene responsible for a disease often identifies polymorphisms either within or close (several kb) to the gene. Polymorphisms such as these mark the gene with the defect responsible for the disorder. The possibility that marker and gene defect can be separated by meiotic recombination, although extremely small, has to be considered. The probability of a crossover event occurring in the human genome is 10^{-8} per nucleotide per gamete. Thus, if the mutation and polymorphic marker are separated by an interval of 3 kb the probability of crossover is 3×10^{-5} per gamete. Mapping by positional cloning identifies polymorphisms associated (at the Mb level) with a specific gene and have on occasion, accepting the probability of recombination (1% cM^{-1}; see p. 160), proven useful in diagnosis.

An intragenic polymorphism is much less likely to be separated by recombination from the specific gene defect it is tracking, thus making it considerably more reliable diagnostically. The heterozygosity of the marker is of importance in order to assess its potential usefulness in determining carrier status. A bi-allelic RFLP is most useful when the frequency of heterozygosity for the allele is 50%. STRs tend to be multi-allelic and have a high frequency of heterozygosity. Multiple intragenic polymorphisms will increase diagnostic reliability and if they are not in allelic association (linkage disequilibrium) will also increase the number of informative carriers. Two (or more) intragenic multi-allelic markers will, by phased pedigree analysis, be considerably more informative than any number of intragenic RFLPs.

8.7 Germ line versus somatic mutation

The possibility exists that the rate of germ line mutation which occurs in humans is at an optimal level sufficient to maintain adequate biological diversity. A higher than required rate of mutation will be detrimental to the overall health of the species since for each beneficial mutation that occurs there are numerous disadvantageous mutations that cause disease. Alternatively, a low rate of mutation will, in the short term, lead to an improvement in the health of the species, but the species risks extinction because of the insufficient genetic diversity available with which to adapt to the many environmental changes that occur in the course of evolution.

Cancer (representing the somatic mechanism of mutagenesis) may play a role in maintaining the optimal rate of mutation within a species by mediating in the selection process. As multiple mutations are necessary to cause cancer, this serves as a means to amplify small differences in the mutation rate. The small number of humans that develop cancer fall into three categories. First, subjects who have a normal rate of germ line mutation but unluckily also mutate tumour repressor genes and oncogenes; second, individuals who inherit a defect in a tumour suppresser gene (or oncogene) (e.g. Li–Fraumeni syndrome or familial retinoblastoma); and, last, subjects with defects in any one of hundreds of genes that affect the rate of germ line mutation.

Index